A NEW PLACE FOR LEARNING SCIENCE

STARTING & RUNNING A SCIENCE CENTER

SHEILA GRINELL

ASSOCIATION OF SCIENCE-TECHNOLOGY CENTERS

Copyright 1992, Association of Science-Technology Centers
ISBN: 0-944040-30-6

Publication of this book was supported by the National Science
Foundation under Grant No. MDR-9150141. Any opinions,
findings, and conclusions or recommendations expressed in
this material are those of the authors and do not necessarily
reflect the views of the National Science Foundation.

For more information, contact:
Association of Science-Technology Centers
1025 Vermont Avenue, NW, Suite 500
Washington, DC 20005-3516
202/783-7200   Fax: 202/783-7207

Design by Hasten & Hunt Graphic Design, Inc.

CONTENTS

AUTHORS OF VIEWPOINTS

G. Donald Adams
Director of Public Affairs
Henry Ford Museum and
    Greenfield Village
Dearborn, Michigan

John Boatright
Group Vice President
The Martin Agency
Richmond, Virginia

Michelene T.H. Chi
Associate Professor of Psychology
University of Pittsburgh
Pittsburgh, Pennsylvania

Peter E. Drucker
Professor of Social Sciences and
    Management
Claremont Graduate School
Claremont, California

David Hawkins
Professor of Philosophy, Emeritus
University of Colorado
Boulder, Colorado

Marilyn G. Hood
Museum Consultant
Columbus, Ohio

Lauren B. Resnick
Professor of Psychology and
    Education
University of Pittsburgh
Pittsburgh, Pennsylvania

George W. Tressel
Communication Consultant
Corporate Advisor, Children's
    Television Workshop
Potomac, Maryland

John Ziman
Director
Science Policy Support Group
London, England

AUTHORS OF CASE STUDIES

Elsa Feher
Director of Science Center
Reuben H. Fleet Space Theater and
    Science Center
San Diego, California

Michael Gore
Foundation Director
Questacon—The National Science
    & Technology Centre
Canberra, Australia

Nils Hornstrup
Director of Exhibition
    Development
Eksperimentarium
Hellerup, Denmark

Charles H. Howarth, Jr.
President
Liberty Science Center
Jersey City, New Jersey

Marcia R. Howe
Executive Director
Carter House Natural Science
    Museum
Redding, California

Janet Johnson
Director of Education
Cranbrook Institute of Science
Bloomfield Hills, Michigan

Karen Johnson
Executive Director
Discovery Museum of Orange
    County
Santa Ana, California

Suzanne LeBlanc
Executive Director
Lied Discovery Children's
    Museum
Las Vegas, Nevada

David A. Ucko
President
Kansas City Museum
Kansas City, Missouri

Cynthia Yao
Executive Director
Ann Arbor Hands-on Museum
Ann Arbor, Michigan

PREFACE

Over the past 30 years, as the number of science centers has grown, a significant body of experience has accumulated about the process of starting and running these lively centers for science learning. Each story is different, and many approaches have resulted in equally healthy organizations. There is no one right way to start a science center.

In commissioning this book, ASTC asked author Sheila Grinell to tap these rich and varied experiences, providing viewpoints and case studies to guide the founders of the next generation of science centers. Sheila Grinell helped start two centers—the Exploratorium, where she was co-director for exhibits and programs from 1969 to 1974, and the New York Hall of Science, where she served as associate director from 1984 to 1987—and served as ASTC's executive director from 1980 to 1981. As a consultant and coordinator of ASTC's Institutes for New Science Centers, she has guided many colleagues through the process of starting new institutions.

In compiling this book, Sheila drew on her own experiences as well as those of colleagues in and out of science centers, including psychologists, teachers, marketing specialists, and students of public understanding of science. The field's tradition of sharing and collaboration, which so generously extends to newcomers, has resulted in the case studies that appear here.

We would like to thank these many contributors, as well as those who reviewed the manuscript and offered helpful suggestions—in particular, Olivia Diaz, Sally Duensing, Al Read, and Dennis Schatz. Susan McCormick, Wendy Pollock, and Jill Reiss of the ASTC staff were responsible for editing the volume.

We also want to recognize the long-standing leadership of the National Science Foundation in supporting development of exemplary exhibits and programs and their wide dissemination throughout the museum field, in encouraging collaboration among museums, and in making possible professional development efforts—including this book— that enrich the quality of informal science in communities throughout the United States, and the world.

Starting any new institution can be a challenge. One founder compared his experience to the loneliness of the long-distance runner—fulfilling, but fraught with challenges. This book should make it easier to finish the course.

Bonnie VanDorn
Executive Director
Association of Science-Technology Centers

ABOUT SCIENCE CENTERS—AND THIS BOOK

For the past several decades, people all over the world have been starting science centers. There are now about 300 of these new-style science museums, two-thirds of them in North America. Interest in establishing additional science centers continues to be strong.

In the towns and cities that are home to science centers, visitors have become accustomed to a museum environment that is designed to be explored actively, sometimes noisily, and in which displays require tweaking and probing to reveal their contents. The traditional definition of the museum—taxonomic collections behind glass cases, touchable only by scholars and conservators—has been expanded.

A family visiting a science center might find exhibits on gravity, levers, and pulleys; on optics and visual perception; on local ecology; and on the human body. As parents and children handle the displays, they act as both experimenters and experimental subjects. They also watch demonstrations (of the dissection of a cow's eye or the effect of the extreme cold of liquid nitrogen on living tissue, for example) and ask questions about the dramatic and intriguing things they see. They enjoy the time spent in a safe, upbeat environment where their curiosity is both stimulated and rewarded.

Today, in the United States, people in smaller communities are founding smaller, privately financed centers. Meanwhile, some of the older, traditional science museums in major cities are expanding and adopting educational goals and interactive techniques characteristic of science centers. Outside the U.S., there is similar diversity, but there tend to be more large, government-supported science centers founded mainly in capital cities. Both large and small centers offer visitors experiences that are similar at the core.

ROOTS

Science centers share a mission and a philosophy. Their mission, broadly speaking, is to help familiarize members of the public with the objects and ideas of science and technology; their philosophy is that learning flows most effectively from situations that foster active participation in handling, observing, and asking questions about artifacts and phenomena.

Inspiration for the development of science centers came from Munich's Deutsches Museum, founded in the early 1900s, on the heels of a succession of celebratory world's fairs. The Deutsches Museum exhibited historical collections of scientific apparatus and industrial machinery, but it also set out to explain their workings and significance to a broad audience. Machines moved, demonstrators explained principles, and visitors handled apparatus. The museum's populist, educational mission and lively techniques were soon emulated in Europe and the U.S., notably in Chicago and Philadelphia.[1]

In the late 1960s, after the decade of reform in science education that followed Sputnik's launch in 1957, several institutions opened that further elaborated on the concept of interactivity. The Exploratorium in San Francisco and the Ontario Science Centre near Toronto eschewed historical and

industrial collections in favor of apparatus and programs designed to communicate basic science in terms readily accessible to visitors. These institutions postulated that displays and programs carefully designed to provide first-hand experience with phenomena could captivate ordinary people and, in the best of circumstances, stimulate original thinking about science.

Since the 1960s, the educational philosophy and methods of the Exploratorium, the Ontario Science Centre, and the half dozen other pioneer science centers have been widely emulated. Science centers and like-minded museums are thriving: Attendance at 130 institutions in 1986 exceeded 50 million, and the overwhelming majority operated in the black.[2]

New centers have built and continue to build on the success of their predecessors. They often present exhibits that have been tested elsewhere and have proven involving and exciting enough to bring people back for more. But new centers also seek to contribute to the repertoire of the field. They build exhibits on new subjects (environmental science and advanced technologies, for example), and they invent programs to serve specific audiences in their locales.

ISSUES ADDRESSED IN THIS BOOK

Regardless of an institution's size or system of governance, people starting a science center—or adding a science center to an existing institution—face the same set of tasks: defining the roles the center will play in the community; developing an audience and understanding its needs; creating exhibits and programs that inspire and support exploratory behavior on the part of visitors and make science delightful and appealing; and running the center on a financially sound and stable footing. Each of these tasks affects the others, feeding back information that reshapes the emerging institution in its gestational years.

This volume addresses these central tasks in four separate sections. Each section begins with an overview that discusses key questions asked by museum founders (or of them by their constituents). Viewpoints by leading analysts are represented in reprinted essays that follow; and brief case studies by museum practitioners translate theory into practice. You may wish to read all four of the overviews, then return to the viewpoints and case studies for a more detailed look. If you want to learn more, you'll find bibliographic information in each overview and at the end of the book.

These chapters don't offer formulas or predictive models of how science centers grow. Building an institution is too complex a process to codify, depending as it does on individual talents, local resources, and timing. But developers will find examples of thought and action others have utilized in creating their museums—examples that may prove helpful in starting a new place for learning science.

Notes

1. For a history of science centers, see Victor J. Danilov, *Science and Technology Centers.*

2. For a survey of the field of science centers, see Susan McCormick, ed., *The ASTC Science Center Survey: Administration and Finance Report.*

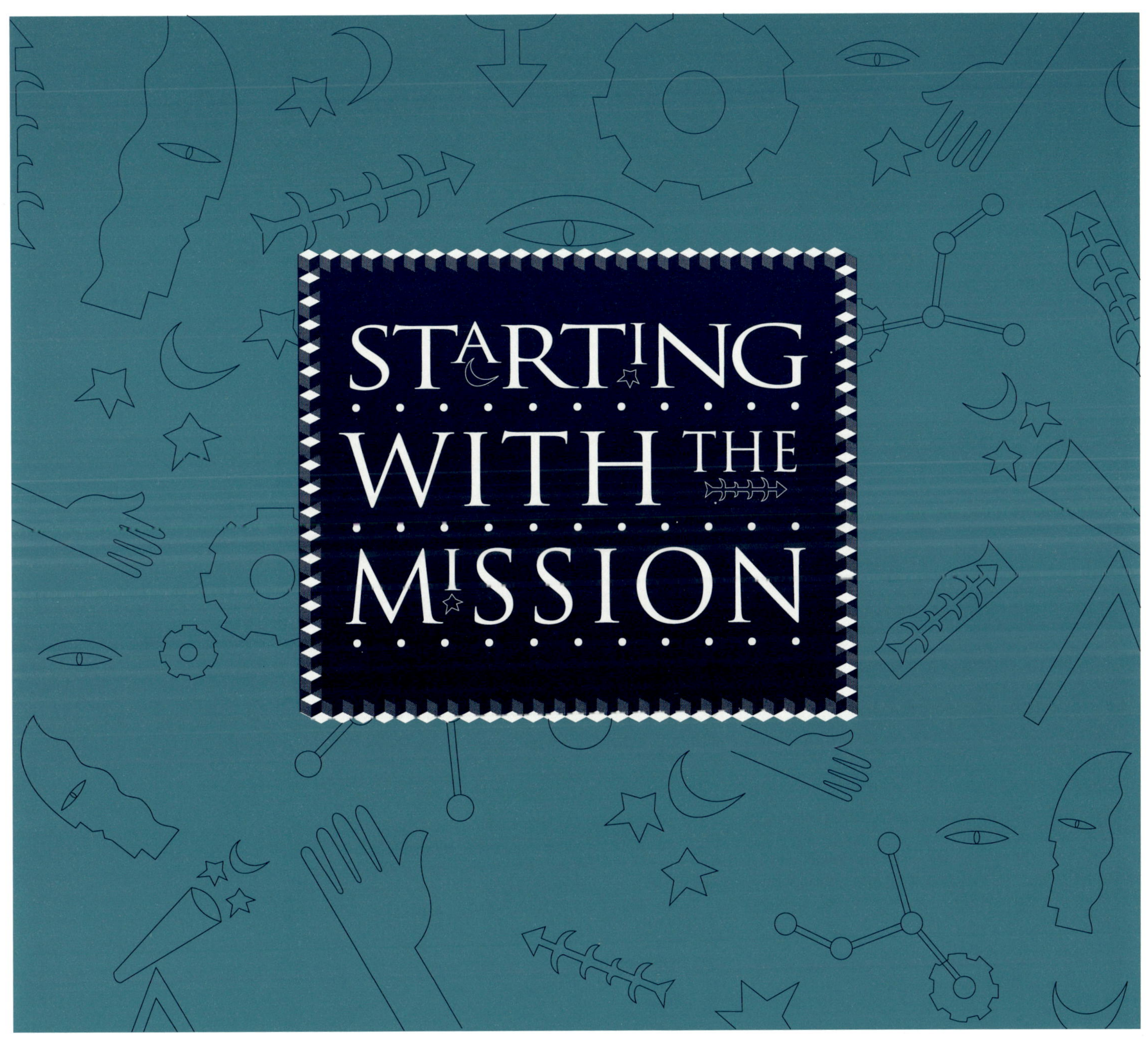

STARTING WITH THE MISSION

People hear about science and technology in a variety of ways. When they are young, they go to school; later, the workplace may provide opportunities to improve their knowledge of the technical basis of their business. All along, they may read about science in newspapers and magazines; watch it on TV; go to museums, science centers, and zoos; participate in hobbies or clubs for amateur science; and talk to family and friends. Although we know a fair amount about how different people use scientific information in specific contexts, no one really knows how these streams of information influence people's attitudes toward and knowledge of science and technology.

In the last decade, government agencies in the U.S. have introduced a new term to describe learning about science outside school (the dominant mode for most people over age 15). "Informal science education," they say, happens through channels that include television, newspapers, advocacy organizations, and exhibiting institutions. Each of these channels reaches a different audience and presents a different kind of information; in most cases, commercial and political interests come into play. A number of reports, committees, and commissions have suggested that informal science education, however fragmented and subject to institutional constraints, plays a significant role in shaping public awareness of science and technology.[1] Science centers and other proponents of interactive science learning have been recognized as important players in this arena.

CONCERN ABOUT SCIENCE EDUCATION

In many countries, leaders have reviewed the status of public understanding of science and technology and found it wanting. Britain's Royal Society issued a report in 1985 that pointed to a disturbing discrepancy: Many people, regardless of socioeconomic status, do not perceive science to be part of their culture, although science and technology play a major role in daily life.[2] In Britain and elsewhere, leaders recognize this as a critical problem that affects many aspects of the social fabric.

Concern in the U.S. has been fueled by a variety of assessments of how American young people compare to their predecessors and to their peers in other countries at various stages of the formal education process. American high school students rank near or at the bottom of multi-nation lists in math, physics, chemistry, and biology.[3] Graduate schools in engineering admit high percentages of foreign students to train for jobs that are seen as vital to sustaining America's leading position in research and development. And girls and minorities, who will form an ever-increasing percentage of the workforce in future generations, are vastly underrepresented in technical education.[4]

Some analysts point to deficiencies in the school system. Very little science is taught in elementary grades, when children may be most receptive, but many teachers are not well-qualified to lead them. In junior and senior high, because students opt out of science, and districts reshuffle the teaching load, a quarter of U.S. teachers wind up teaching courses for which they are not certified.[5] Although teachers prefer laboratory lessons, they rely on textbooks and lectures—textbook lessons take less time, money, and resources to prepare. Yet, many educators fear that such lessons turn into vocabulary exercises that do not portray science as an active—or attractive—pursuit of knowledge. Several comprehensive plans for reform are now beginning to be implemented at test sites across the United States to improve what is taught to students, and how and when.[6]

But reformers recognize that schools alone can't ensure that an increasing number of youngsters will be prepared to learn science, or even motivated to try. Students' attitudes and choices are influenced by family, friends, media, and the possibilities revealed by their own experience. They need many varied and comprehensible encounters with science in order to want more of it. Reinforcement must come again and again, in a variety of ways. Each of the informal science education channels can help provide reinforcement, reaching a large audience with its own kind of message. Governmental leaders have called for all sectors of society to use these channels, and mobilize new resources, to help provide a social context that favors science learning—for youth and adults alike.[7]

Schools alone can't ensure that an increasing number of youngsters will be prepared to learn science, or even motivated to try.

For over a decade, political scientist Jon Miller has been surveying adult Americans to determine their attitudes toward and knowledge of science and technology. Miller's work has been paralleled in the United Kingdom by John Durant and colleagues at Oxford University, in Japan, and to a lesser degree in several other industrialized nations. All these surveys produce results that are substantially alike, although specifics may vary.[8]

Miller and Durant found that the American and British peoples express strong interest in and support for science and technology, in general. They also feel fairly well-informed. But their responses to factual questions belie their self-reporting. Using a three-part measure of "science literacy" (understanding of the processes of science, basic vocabulary of facts, and understanding of the role science and technology play in

society), Miller says that only five to six percent of adult Americans are scientifically literate. While other analysts question the way Miller has structured this measurement, his work has raised widespread concern about the science educational needs of adult populations in an increasingly high-tech world.

A ROLE FOR SCIENCE CENTERS

Scientists and engineers share a culture, with a sophisticated language, articulated mores, and strenuous initiation procedures. The rest of us depend upon the products of that complex and sometimes fractious culture in our daily lives. Most of us are novices in science and technology (perhaps reluctant novices), and our ability to use and interpret scientific culture for ourselves is more or less limited. Science centers can help narrow the distance between expert and novice in science and technology.

Science centers provide *access* to science and scientists, they improve our *comfort* level with technical ideas and artifacts, and they enable us to obtain first-hand *experience* with phenomena. Operating in three dimensions, unlike print or electronic media, science centers can enable visitors to handle materials and to question the experts or their interpreters face to face. This kind of direct contact provides powerful motivation for visitors to get more involved.

Access

The science center is a destination, a place to go, one of the few places anyone can see natural rarities or ask a question about science—and where one is encouraged to ask. The center gives science a local presence and a face that might not otherwise exist. Center staff invent ways to bring science, citizens, and students together, including "camp-ins,"

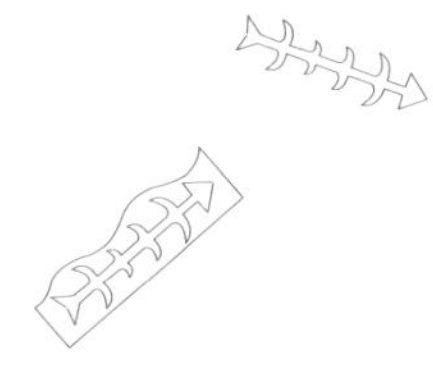

planet-viewing festivals that feature real-time transmissions from satellites, and "Sunday afternoon conversations with scientists" complete with dramatic props, to name a few of their innovations.

Most science centers are more flexible and entrepreneurial in structure than the local school systems. They are able to launch ventures and make alliances for science learning that schools cannot. Starting a pilot program with only a small financial investment, science centers can serve as efficient incubators for improving a community's science education resources. As long as the science is solid and the access is easy, this kind of community-oriented service can flourish in a wide variety of forms.

Science centers can serve as efficient incubators for improving a community's science education resources.

Comfort
Science centers allow visitors to look at science and technology without the formal prerequisites that are intimidating to so many. Visitors don't need a background in mathematics or prior knowledge of the subject matter. There are no grades, and no one has to go it alone. Centers seek to increase visitors' level of comfort with science and technology by offering opportunities for visitors to succeed in their approach to the material. When they present phenomena and interpret ideas in clear, non-threatening, aesthetically pleasing, and physically engaging ways, a visit turns into a relaxing exploration—instead of a show of ignorance—and great fun.

Most science centers encourage visitors to exercise personal choice in delving into the material. They expect visitors to approach exhibits idiosyncratically and do not organize content in linear story lines. They also expect visitors to use displays in family or peer groups, and they try to include something for everyone. Research shows that personal choice and cooperative exploration enhance science learning.[9] They certainly make for an enjoyable time, one that families will want to repeat.

Experience
Philip Morrison and others have pointed out that people today, growing up in urban settings and absorbing information passively from television, do not labor with ropes and levers and fire and liquids as farmers and artisans did, and so do not have well-formed intuitions about how the natural world behaves.[10] Schooling teaches us rules about this behavior, but not about the real-world practice to which the rules apply.

In order to appreciate the power and elegance of a rule, a student has to have some sense of why it is needed, of the many odd things happening in the physical world that it might explain. Students have to play around with reflections and images—to see things where they're not supposed to be, or laugh at comical distortions, or feel invisible infrared focused on their hands—in order to care about the notion of "focal length." Tinkering with mirrors and lenses will build their interest and confidence; they will have solid ground from which to develop their own ideas.

Science centers are dedicated to providing opportunities for students (and parents and teachers) to play around first-hand with materials and experimental apparatus, and so to build intuitions about the natural world. They may also seek to convey information about a particular topic or to familiarize visitors with the habits of mind scientists use to develop knowledge about nature. But in any case, they distinguish between their role and that of the schools.

Science centers don't expect visitors to approach their offerings systematically—as if reading a book or taking a course. They expect most visitors to browse, directing their attention where they will. Through such episodic encounters with engaging material, science centers hope to lure, stimulate, and invite visitors to discover something new—just one thing—about the structure of the physical world. They seek to empower visitors to take one step closer to the expert's point of view with every visit, recognizing that each individual visitor, depending on background, interests, and talents, will take a different step.

ASSESSING ACCOMPLISHMENTS

Science centers are committed to helping visitors make their own discoveries. Formal science educators also value teaching methods that promote student inquiry leading to discovery. But exactly how inquiry and discovery should be construed or assessed in school settings remains the subject of debate.[11] Evaluators disagree about methods for measuring discovery in schools, where populations are controlled and testable over time. In science centers, where there are no exams, where the visitor population is so varied, where there may not be repeated contact with a visitor, and where educational goals are so broadly defined, assessment is even more difficult.

The really tough question, though—and one frequently asked by science centers' supporters—is about the impact experiences at a science center have on how much science people learn. No one has come up with a quantitative way to document the effect science centers have on the visitor population. Part of the problem is that we can't predict when learning will happen. A child may see something in an exhibit that enables him, months later, to grasp a concept the first time it is taught at school. Nor can we anticipate the aesthetic or social value of the experience for individual visitors. Frank Oppenheimer, founder of the Exploratorium, used to talk about a woman who reported that visiting that center gave her the confidence she needed to rewire a lamp. How could you capture such an outcome in a survey or interview at the exit door?

Several studies have measured the effects of science center programs on special audiences, like teachers in a training program or students working as floor staff. The Exploratorium surveyed adolescents who had worked as "explainers" during a 10-year period and found that their experience had influenced them in lasting, positive ways.[12] A brief stint as an explainer (typically six months) improved teenagers' communication skills and increased their interest in learning science. Few intervention programs with adolescents do as well, as quickly. While such specific studies can help shape policy at science centers, the question of overall impact remains unanswered.

In the absence of hard numbers, science center staff and supporters turn to anecdotes and observations of the life on the exhibit floor. They sense that something valuable is happening to visitors that could not readily happen elsewhere. They see that the science center environment motivates visitors to explore technical subjects. They feel, with some support from the research literature,[13] that this exploration lays groundwork for the development of scientific understanding. Clearly, their conclusions are shared by people in many communities who are planning science centers of their own.

DEFINING A SCIENCE CENTER

Within this framework of common purpose, science centers choose different agendas. There are many variations on the

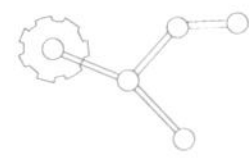

science center theme, each with its own particular character.

Some centers elect to serve the entire population while others focus on special constituencies. One small center concentrates on the local elementary schools, assuming formal education functions that the school district cannot perform. Another center affiliated with a university specializes in pre-service training for undergraduates interested in science teaching as a career. Another center in an inner-city neighborhood works extensively with church and civic groups to involve families that otherwise might not visit. Other centers find ways to communicate over long distances with people in remote areas. Centers in developing countries sometimes bring practical technical assistance to the poorest communities, such as teaching people to test their drinking water, and explaining why they should.

Some science centers claim all of science as their territory, while others concentrate on a particular topic in science, medicine, or technology. They may own a remarkable artifact or they may develop exhibits and programs on their region's major industry—for example, nickel mining in Sudbury, Ontario. Some centers focus on the special geographical features of their locales. Others develop exhibits and programs on subjects that staff and supporters find fascinating, or in which they have special competence. A scientist from Fermilab built exhibits on particle physics; Parisian educators concentrated on the relations between science and society.

All of these formulations work, but they have no predictive value for new science centers. Nor do the relevant professional associations prescribe what a science center should be. The American Association of Museums accredits science centers on very general criteria that would exclude few emerging institutions. The ASTC Board of Directors has,

however, issued guidelines for all science centers, old and new, to increase their efforts to serve wider audiences, in particular minorities, women, and people with disabilities. (The guidelines appear in Appendix A.) For ASTC and its membership, reaching these audiences, underrepresented in science and often underserved at museums and science centers, has become a priority.

Each new science center has to define its own aims, methods, audience, and content within the context of local possibilities. Planners must investigate resources, consider local needs, and negotiate among supporters about what their institution should do. They can clarify the values of their new institution by writing a concise definition, a statement of mission. It usually takes some time to settle on the language—and it should, because the language and its implications define the course of action they will take in developing the center.

Each new science center has to define its own aims, methods, audience, and content within the context of local possibilities.

A mission statement is more than a formality. The board and staff of a developing science center need guidelines for evaluating whether or not to accept offers of equipment, or even a building that may be costly to renovate. They need a basis for making strategic decisions about which exhibits to build and which programs to implement first. Should they try to produce a TV show to improve visibility in the community or run clubs after school? It depends on what they wish to accomplish—on the definition of purpose and methods they have carefully forged. Finally, they need a yardstick against which to measure their progress. A mission statement prescribes objectives as well as goals.

Often when people are starting from scratch, defining their science center is an incremental process that takes years as various avenues are investigated and options eliminated. Defining an institution requires patience, persistence, and tolerance for ambiguity until the job is done. And by then, it may be time to revisit the mission statement, making adjustments to reflect the institution's growth.

ASSEMBLING RESOURCES

Whatever their specific agendas, science centers need an array of assets to accomplish their general purpose. They need a user-friendly building, filled with exhibits and programs, that is maintained, insured, cleaned, and kept open on weekends and holidays. They need a skilled staff to supervise operations, interact with visitors, and plan new offerings. They need effective trustees to review policy and obtain political and financial support for the long term. And they need an audience to come, enjoy themselves, and get something out of the visit—access, comfort, and experience with science.

No one starts a science center with any of these in hand. People begin with a sense of the value a science center can bring to their community, and they look for like-minded souls. While the paths they take will differ, they eventually need to tap four resource streams: scientific, educational, political, and community. Without grounding in each of these areas, a science center is not likely to thrive. People from all four areas give it strength and develop the necessary material support.

To hold the visitor's attention, a science center needs to present phenomena worthy of display and ideas worthy of investigation. It needs to convey the excitement about science a committed professional feels, and to do so in a balanced and accurate fashion. A high-quality presentation can only be designed with the extensive participation of a person or persons who know science intimately and can share its wonders. But without interpretive techniques that make phenomena tangible or familiar language to discuss abstract ideas, the visitor gets lost. Planning also requires educators sensitive to the audience's intellectual needs and skilled in popular communications.

Most science centers rely on volunteers to get started, and funding contributed by private and public sources. To ensure an adequate supply of both, science center planners need to develop political support among regional decision-makers. Sometimes political support comes after the project achieves a degree of popular recognition—through demonstration exhibits or outreach programs, for example. Planners need to involve community groups in their own right, though, to cultivate an audience and obtain feedback on how welcome that audience will feel when the institution eventually opens.

Part of the early work in starting a science center is establishing contact with organizations, companies, and individual leaders who might help in scientific, educational, political, and community circles. Through conversations with these people, planners can identify resources to tap in building their center. Planners should not forget to involve nearby museums, science centers, and other cultural institutions. These groups already enjoy political support, and their approval, or lack of it, can be critical to the success of a new project.

New science centers grow slowly, with pieces falling into place in no particular order. Gestation times extend from several years to well over a decade. So far, there have been few outright failures. The people who have started science centers have worked very hard, learning a wider range of skills and facing a wider range of issues than they could possibly have imagined at the outset. It's been a long haul, but an exciting one.

Notes

1. See chapters on informal science education in Michael Knapp et al, *Opportunities for Strategic Investment in K-12 Science Education*; and Office of Technology Assessment, *Educating Scientists and Engineers: Grade School to Grad School.*

2. The Royal Society, "The Public Understanding of Science."

3. For performance of U.S. students, see Educational Testing Service, *A World of Differences*; and T. Neville Postlethwaite and David E. Wiley, *The IEA Study of Science II: Science Achievement in Twenty-Three Countries.*

4. See, for example, Shirley Malcom, "Who Will Do Science in the Next Century?"

5. Iris Weiss, *Report of the 1985-86 National Survey of Science and Mathematics Education.*

6. Two important reform efforts are: Project 2061, sponsored by the American Association for the Advancement of Science; and Reform of Scope and Sequence, sponsored by the National Science Teachers Association. Both organizations are located in Washington, DC.

7. The Bush Administration's "America 2000" education strategy, conceived with the nation's governors, and the U.S. Public Health Service's "Prologue to Action" strategy call for increased involvement of non-school, community-based organizations.

8. Jon Miller, "Scientific Literacy"; John Durant et al, "The Public Understanding of Science."

9. See, for example, Roger Johnson and David Johnson, "Cooperative Learning and the Achievement and Socialization Crises in Science and Mathematics Classrooms."

10. Philip Morrison and Phylis Morrison, "...But TV is Not Enough."

11. For discussions of inquiry methods, see Eleanor Duckworth, *The Having of Wonderful Ideas & Other Essays on Teaching and Learning*; or Joan Solomon, *Teaching Children in the Laboratory.*

12. Judy Diamond et al, "The Exploratorium's Explainer Program: The Long-Term Impacts on Teenagers of Teaching Science to the Public."

13. See Beverly Serrell, ed., *What Research Says about Learning in Science Museums.*

V I E W P O I N T S

The following articles lend perspective to the basic mission of science centers: promoting public understanding of science. Former National Science Foundation division director George Tressel explains how formal and informal science education interrelate, and describes the role of science centers and museums within this framework. John Ziman, summarizing the results of a British research program on science communication, defines a new model for thinking about audiences for science, which takes into account the social and personal contexts in which people acquire and use technical information.

THE ROLE OF INFORMAL LEARNING IN SCIENCE EDUCATION
George W. Tressel

In many ways it's easier to talk about schools than it is to talk about informal learning—as if informal learning were the same thing, only with exhibits. After all, we all went to school, and so we are all experts with quick cheap answers to incredibly complex and deep-seated problems. Would you like to hear a few? Here are some problems and quick fixes that I stole from an article in an Iowa newspaper.

Problem: Teachers are burned out. They teach for endless hours with little or no time for preparation—so much so that they find it difficult to get up and go to school or to stay there after they get there.

Magic Solution: Lengthen the school day and the school year without adding teachers.

Problem: Large numbers of high school students can't read, write, or think.

Magic Solution: Make them take more math and science—and give them tougher tests.

Problem: Teachers suffer from low morale. They are blamed for everything and are depressed because they see themselves paid less than many unskilled and semi-skilled workers.

Magic Solution: Give a handful of teachers more money so the others have a new basis of comparison.

I am too old to believe in such magic solutions. Someone once said that "there is no problem so difficult or so complex that there is not a simple, obvious, and direct solution—that is also *wrong!*"

Let me get to the point. Schooling is what happens in schools: training to provide skills. Skills to use in careers, skills to use in life, skills in preparation for a lifetime of personal, self-directed informal learning.

Because, in fact, most people, most of the time, learn most of what they know *outside* of school; one of the principal roles of formal education is to provide skills and background that make this possible.

The question of preparation is the salient relationship between formal and informal learning. The process of informal learning is wrapped around our formal education. It provides preparation and support for schooling—which in turn provides skills and preparation for lifelong informal education.

This relation is important because each stage reflects the success or failure of the previous stage. You can only learn what you almost know already.

And you are certainly not likely to become an enthusiastic, successful lifelong learner if your previous experiences have been painful, dull, and characterized by failure. Let us be sure that we do not forget this. The most important aspect of informal learning must always remain. It is fun.

In our studies of why children watch "3-2-1 Contact" and "Square One TV," we found they all learn different things and they go away doing different things. Part of the fun is this freedom—informal learning should complement school, but it should not *be* school.

WHY WE NEED INFORMAL EDUCATION

Let me turn a little bit to the reality of what our school systems look like and why we need this sort of thing. This is the pattern of what I like to call mainline education (Figure 1). In coarse terms, most of our kids finish elementary school. But by the end of four years of high school, 25 percent have dropped out, a little more in some cities, a little less in others: 25 percent of the entire population.

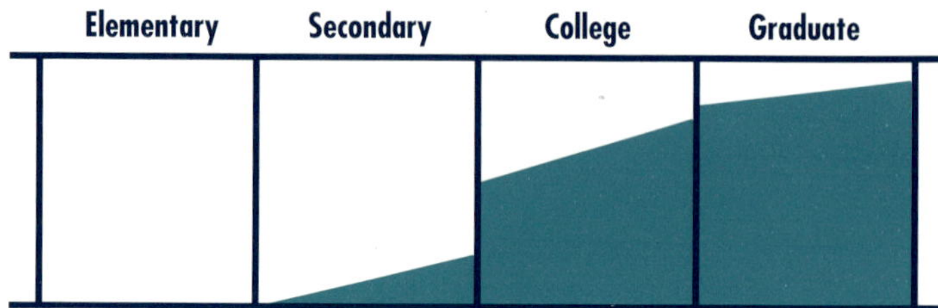

Figure 1. The Education Flow Pattern

Some 50 percent go on to enter college, and roughly 25 percent finish four years. About 5 percent go on into higher education, and again there is some attrition.

This is the mainline flow through our education system, and as you can see, for all practical purposes the population is divided into two parts. Fifty percent come out of the system during or at the end of high school: the *general workforce.* These are the people who will service our cars, dish up hamburgers, and run our assembly lines. The general workforce: 50 percent of the population, half of whom did not finish high school.

The other 50 percent, the *professional workforce* who come out during or at the end of four years of college, include the people who are going to be our legislators, doctors, lawyers, writers, managers, and investors. These are the people who will decide whether or not to spend money on the supercollider, whether or not to invest money in new steel mills and new technology. Finally at the very end are the 5 percent who become all of our so-called intellectual resources: our academics, engineers, research scientists, and so on.

Within this mainline is the "pipeline" (Figure 2) that produces our technological workforce. This includes almost all of our scientists, engineers, technicians, and designers.

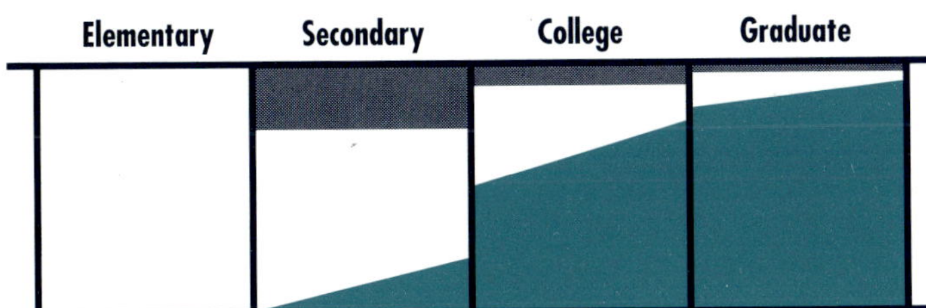

Figure 2. The Science Education Pipeline

What do we teach them about science in elementary school? Nothing. For all practical purposes we do not teach any science in elementary schools. One hour a week of so-called science does not count. We wait until middle school or high school, where we ask a 14-year-old the magic question: "Would you like to take physics, chemistry, botany, and zoology?"

The answer is usually "No." Only 23 percent of the total population say "yes"—defined as three years of math/science. Typically that is some algebra, some general science, perhaps some general biology, botany, or zoology. And in very rare circumstances a course in physics or chemistry.

Most students who do enroll in these courses do so only because it's the college entrance requirement. Not because they want to be an engineer or physicist or what have you. Roughly, only one out of five are taking a disciplinary course because they mean to pursue that area. The others are taking it because they want to go to college.

And in fact that's what happens when they get to college. Five percent of the total population go into the science and engineering fields. Attrition cuts that roughly in half by the end of four years; and then less than one percent, 0.8 percent of the population, go on into graduate study. Here in this minuscule tail is where the Nobel Prize winners come from, and where all of the elite of our research scientists and graduate engineers must come from.

NOW THEN, WHAT DOES THIS MEAN?

It means that in high school we are giving most college-bound students a couple of years of disciplinary science courses. Most of them have no disciplinary interest at all and will take little or no further science in college. Is that what our decision-making professional workforce needs? Are a couple of years of high school botany and zoology the right preparation for a decision-maker who is going to decide about bioengineering?

Is this what newspaper writers should have for preparation? Is this what our lawyers and our legislators need? Do people like this need disciplinary courses? In effect, we are saying that a couple of courses in, say, general science, botany, and zoology provide a suitable overview of science for a typical decision-maker. We are using disciplinary courses as a substitute for an integrated view of science and technology—and that doesn't make sense.

At the same time, this means that the student who does want to be a chemist is sitting in a classroom with four other kids who aren't really interested in chemistry as a career; they're only taking it because they want to get into college. That is certainly no help to the one who has already chosen a profession, nor is it the way to raise our standards of professional preparation.

Thus, we are serving neither the general nor the pre-professional student well. Further, if we are concerned about the supply of technical students in college, we must face the fact that we are already drawing a substantial portion of the potential; to draw more we need to address the supply.

The salient problem occurred years before, when we asked a 14-year-old the magic question: "Do you want to take math and science?" Who said yes? My kid said yes. Your kids said yes. Our kids said yes because 1) they were well prepared, and 2) we told them to say yes. My kid is now 42 years old and he is a biochemist. When he was nine years old he had a camera, a microscope, a telescope, and a father who took him to the museum. People like us do something outside the system that isn't being done inside the system: We provide a very rich environment, and we prepare our children for success.

Let's face it: Our kids were lucky enough to pick the right parents. Parents who buy them books, cameras, microscopes, and take them to the museum. Parents who ask them, "What did you watch on the tube yesterday?" We made sure they were prepared and had every chance of success. We make life as rich for them as possible. We tell them they will succeed. And they do.

We are going to have to change what happens in school and out of school.

Science education is inherently elitist, because it is a meritocracy and is intensely competitive. That is well and good for anyone who has a fair start. But the system is stacked. It is stacked, first of all, against people who aren't socially affluent. Second, it is stacked against women, minorities, and people with disabilities. Because by and large we tend to convey stereotypes of what people are going to be when they grow up.

We give little boys things to break, things that you can get hurt with, things that you can tear up and pound on, and then fix. And we don't give them to girls because "you might get hurt and it's not nice." We don't mean to do this, but much of our education system serves as a selection system and a filter. And when you listen closely, many of the "quick fixes" for our educational system are really asking for more of this. Don't educate them better: Raise the standards, filter harder.

Another "solution" is to recruit more. Get more of them to go into science and engineering in college. But you already recruit roughly one of five from the college pool. No one can object to recruiting or higher standards. But unless you build the pool, you will not win. The system needs to change in two ways. We are going to have to change what happens in school and out of school, in the early years before the filtering question is posed.

Outside of the system, we need to make a richer environment for more kids. We need to make it as rich as we can in every possible way. This is easier if the kids have wonderful parents who lean over their shoulder and encourage them on to a rewarding career. But a lot of us don't. And we've got to figure out what to do about that. We have more and more single-parent homes, and it's not enough to say it's the parent's problem; it's also society's problem.

Second, we need to put into place a consistent, coherent pattern of K-12 science education for all students, and it should serve three purposes:

- First, it should serve the 50 percent who will leave the system at the end of high school. Give them enough math/science background so they have a realistic ability to be trained for technologically related jobs. Enough so they can read about science in the paper at more than the fourth-grade level.

- Second, this curriculum should lay a groundwork for disciplinary courses so that students are able to make a meaningful choice, and realistically able to succeed if they do so.

- Third, this curriculum should provide a really comprehensive view of science before students go into college so that we produce doctors, lawyers, and decision-makers who have a fluent perspective of science, not simply a course in chemistry and a course in biology.

This is not all that is needed. We need an overall system where every student has opportunity, encouragement, and challenge at every level of interest and ability. This means that this basic pattern for all students should be paralleled in high school with the best pre-professional courses we can design, taught by the best teachers we can provide, at the highest standards in the world.

And this in turn should be paralleled by special programs for the most bright and talented: the next generation of Nobel Prize winners. There is something special in the 10-year-old who tests at college level, and we need the best of programs to encourage and develop such talent—especially when we find it among the groups who are least likely to succeed.

THREE STAGES OF EDUCATION

Throughout the K-12 period, our system should parallel the habits of really good parents. Good parents do not test little children every day and tell them they failed. We tell them everything they do is wonderful. We expose them to all of the stimulus we can. We give them a pat on the back and say,

"Gee, are you wonderful, you're really good at that, aren't you! What else are you going to do now?" All you want is for them to keep exploring, and learning, and enjoying the process. Because we know that this is what makes a good, strong person who wants to learn.

The role of the earliest stage is to produce children who like to learn and like to solve problems. We need to embellish and encourage that developing curiosity, both inside and outside of school. Like good parents, we need to give every kind of opportunity—to play the violin, draw pictures of a paramecium, or build a telescope. Like good parents we need to heap praise, so that children learn that performance is what gets us excited.

Then, somewhere along the way, this "no-fail" stage must begin to change, as it does with good parents. Somewhere, after all of their stimulus and reward, good parents become a little harder to please. Somewhere along the way—about middle school—like parents, the system should start asking, "How did you do on that? You ought not to have a 'B'; you ought to have an 'A,' because we know you're really good at that. You need to try a little harder."

We need an overall system where every student has opportunity, encouragement, and challenge at every level of interest and ability.

Good parents unconsciously keep tightening the screw; it gets harder and harder to please your pop. Eventually you internalize this so much that when your parents die you spend the rest of your life still trying to please them anyway. That's what makes good people.

And that's what our system should do:

- It should start with a touchy-feely, success-oriented, no-fail stage where we build a conceptual vocabulary and a habit of solving problems. Enjoy it; don't fail.

- Somewhere in the middle we should be starting to ask, "Let's see, you're going to do an experiment. Why do we want to bother doing that? What do you think you are going to get from this? How are you going to do this experiment? Write it down and tell me afterward what you did." Bit by bit, we must begin to introduce discipline and demand.

- And when we get to the third stage, it should be even more demanding, more relevant, and more content-filled than what our disciplinary courses are right now.

We can build something like this. And this is the pattern of what I believe is needed within the school.

At the same time, there is a major role for informal learning in this process. Setting the stage for school, paralleling it with out-of-school activities, providing first-hand experiences, interaction with curators and research people, and so on.

BUILDING A COHERENT SCIENCE EDUCATION ENVIRONMENT

Among museum people, it's common to think of informal science as what we do. As if museums are the whole show, and they are not. Informal learning takes place through reading, watching TV, libraries, clubs, museums, and a great many other channels.

At the National Science Foundation, we are trying to focus on a coherent pattern of three general kinds of activity. Broadcast-

ing is a way to reach enormous numbers of kids. As most of you know, we now have a daily half-hour TV science program and a daily half-hour math program. More than half of all four to 12-year-olds in the country watch the science program periodically. Thirteen percent of all children watch 40 percent of the programs.

You can walk into a classroom and say, "Does anyone know what *3-2-1 Contact* is?" And all the hands will go up. They can tell you who the characters are. They can tell you about things they saw. That's a lot of science for a lot of kids. We also know that 60 percent of them do something as a direct result of watching it. They read books. They do experiments. They make things. They go with their parents to the museum.

Which is where we came in. Museums are the one institution that understands what this is all about and what we are trying to do—and are doing it already. So at the National Science Foundation, our second layer of priority in informal education is science museums.

Progress in the museum community has been very heartening. The number of science centers has roughly doubled in the last decade. And that coarse number says nothing about the change in the kinds of new activities in museums. Camp-ins for girl scouts. Workshops for teachers. And a million other things that you might not expect in a "museum," like special activities and events for activity groups.

Science activities, especially in museums and established activity groups, are the National Science Foundation's third level of informal education. We are now funding 4-H Clubs, Girl's, Inc., Girl Scouts, and so on, in a concerted effort to give kids direct hands-on experience.

Museums are the ideal place for this to happen. A Girl Scout group leader doesn't need to pretend that she knows an awful lot about science if she takes her troop to the museum. Museums are providing camp-ins, where the leader need only bring the audience and can learn along with them.

We need to continually hone this pattern and remember that the effects of informal learning are gained through time, redundancy, and repetition. We need more attention to what happens after the visit—more means to generate sustained activity and follow-through activities at home.

We need to avoid getting carried away by our own buzz words—words like "relevant" content and "interactive" displays. We need to remember that "relevant" just means important to the person. What's important to me may not be important to you. And for both of us it will change with time.

We need to remember that "interactive" only means I am actively involved; it does not necessarily mean that I push a button.

There is a time when discovering, handling, manipulating is the best way to make things important and the best way to internalize the experience. There is a time when standing alone in a darkened room—meditating on a 9,000-year-old little clay ball containing miniature oil jars (the first form of writing)—is a profoundly involving experience.

We need to avoid dogma; every opinion should not become a school of thought. In the end this is an art, and the only way to be sure what works with a particular audience is to try it, and watch what happens.

- At the simplest level, a person may simply enjoy and remember a few descriptive aspects—familiarity is not a trivial outcome.

- At a higher level, a person may understand the phenomenon or message.

- At a still higher level, a person may be able to extrapolate a concept to other topics and concerns.

These are all worthwhile, and we cannot accomplish them all with every person. We need to separate our goals into the core content that we would like everyone to receive, the added detail that we hope some will receive, and the subtleties that we wish for the most sophisticated parts of the audience.

We cannot do all of these for everyone. We need to stay humble and realistic, and remember how minimal is the average understanding of our audience.

George Tressel is a communication consultant and corporate advisor to the Children's Television Workshop. He serves as an advisor to Quality Education for Minorities, as a board member of Free The Children Trust, and as an advisor to numerous museums, including the Smithsonian Institution and the New England Aquarium. Mr. Tressel is former director of the Materials Development, Research, and Informal Education Division of the National Science Foundation. He has been involved in many aspects of informal education and science communication, and has authored policy and management studies related to broadcasting, exhibits, and outreach activities.

Permission to reprint granted by George Tressel; speech delivered at the Science Learning in the Informal Setting Symposium, November 12-15, 1987, The Chicago Academy of Sciences.

NOT KNOWING, NEEDING TO KNOW, AND WANTING TO KNOW

John Ziman

Any question of measuring progress in public understanding of science has to be asked in relation to prescribed *boundary conditions*. We must decide where this progress is starting from and where it is supposed to be hcaded. Boundary conditions presume a *frame*. There has to be some agreement about what features of the world are *not* going to be discussed or are to be taken as constant. To construct a coherent frame requires a *model*. We have to find an aspect of the world that can be depicted in simple, interrelated concepts.

There are three types of questions that can be asked about public understanding of science. Each is posed and framed according to a different model of science and of its social role. These models are not just slightly different perspectives on the same scene; they are so different in principle that they produce nearly contrary precepts and policies. Let me go through them in turn.

THE DEFICIENCY MODEL

The first of these models is so familiar that it is usually never questioned. According to most scientists, the great majority of ordinary people have very little understanding of science. The outstanding feature of the situation is deemed to be public *ignorance*. The primary question is, "What do people not know—and for goodness' sake, why not?" The problem is perceived as a *deficiency,* which must by all means be overcome. The basic measure of progress in public understanding of science is taken to be how much more science people can be made to understand.

Do we need to go any further? Concern about the knowledge gap between the world of science and the world at large is nothing new. The British and American Associations for the Advancement of Science were founded in the early 19th century to fill this gap. Many scientists of distinction have given of themselves to write, lecture, and broadcast to extend their work to the "attentive public." Many talented writers and broadcasters have made it their business to explain in simple terms what the scientists were doing. Much of the effort reported in workshops on public understanding of science is consciously or unconsciously motivated by this model.

Nevertheless, there has always been regret that these efforts did not seem to be having much effect. This uneasiness came to a head in the report of the Royal Society's Committee on Public Understanding of Science, chaired by Dr. Walter Bodmer, which appeared in 1985.[1] The background to the whole subject was explored very thoroughly, and yet we could find very little serious research on what was going on. The Bodmer report was the cue for the British Economic and Social Research Council, backed up later by the Advisory Board of the Research Councils and by British Gas, to set aside funds for a major program of research. Our task in the Science Policy Support Group was to organize a competition for peer-reviewed proposals, coordinate the work of the various project teams, and make sure that the results of the research were widely disseminated.

Some of the projects were started in early 1987 and are now complete. Other projects started later and are still under way. The scale and diversity of the whole program can be gathered from the project titles (Figure 1). The researchers came from very diverse disciplinary traditions—anthropology, education, history, philosophy, politics, psychology, and sociology, not to mention such interdisciplinary areas as "media studies" and "science, technology, and society." It is worth remarking also that quite a number of them were originally educated to a high level in one or another of the natural sciences.

This diversity of starting points and approaches might have produced a very confused and incoherent outcome. Surprisingly, however, the research audits converged on a few general messages that were not originally obvious and that were, indeed, somewhat contrary to received views. In particular, the research revealed some of the difficulties that arise when one tries to use the deficiency model as a guide to understanding and practical action.

Figure 1. Public Understanding of Science Projects and Principal Investigators (United Kingdom)

1. A National Survey of Public Understanding of Science in Britain: John Durant, Science Museum

2. Young People's Responses to Scientific and Technological Change: Glynis Breakwell, Surrey University

3. Frameworks for Understanding Public Interpretations of Science and Technology: Brian Wynne, Lancaster University

4. Science and the Identification of Strategic Research Areas: Peter Glasner, Bristol Polytechnic

5. Nature's Advocates: The Public Role of Scientists in Conservation Movements: Steve Yearley, Queens University, Belfast

6. Audiences, Disseminators, and the Local Understanding of Technical Issues: Alan Irwin, Manchester University

7. Genetic Disorder: Self Help, Knowledge and Dissemination: Hilary Rose, Bradford University

8. Science Knowledge and Media Influence in Classroom Discussions of Social Issues: Joan Solomon, Oxford University

9. The Production and Presentation of Science in the Mass Media: Anders Hansen, Leicester University

10. Science in the Museum: An Ethnography of a Science Project: Roger Silverstone, Brunel University

11. Primary Teachers' Processing of Scientific Information: Jon Ogborn, Institute of Education (funded by British Gas)

The first difficulty is that "science" is not a well-bounded, coherent entity, capable of being more or less "understood." This finding is not in any sense a subversive attack on the marvelous and immense body of knowledge and practice produced by natural and social scientists, engineers, physicians, and technologists. It is a reminder that what counts as science is defined very differently by different people—or even by the same people under different circumstances.

Scientists themselves do not have a clear and consistent notion of what "science" covers, and often disagree profoundly on what it is telling us about the world. To give some obvious examples, scientists are just as divided as the public at large on whether psychology is a science, they still seem very uncertain about the biological evolution of the human species, and they offer patently contradictory "scientific" advice on healthy

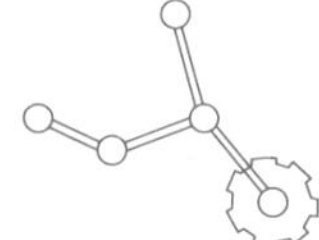

eating. Should our instruments for measuring public understanding of science therefore take no account of knowledge of the social or behavioral sciences? How should they score responses on interesting topics that are still highly disputed? Where would they draw the line between a plausible conjecture and the wisdom of experience?

In other words, "science" is not a special type of knowledge that only starts to be misrepresented and misunderstood outside well-defined boundaries by people who simply do not know any better. The deficiency model, which tries to represent the situation simply in terms of public "ignorance" or scientific "illiteracy," is thus unsound in principle and does not fit the results of our research.

THE RATIONAL CHOICE MODEL

The standard justification for improving public understanding of science is utilitarian. The basic question becomes, "What do people *need* to know in order to be good citizens—even to survive—in a culture largely shaped by science?" If they had no means of fulfilling this need, they might find themselves quite unable to make any sense of important practical questions affecting their lives—small questions like the goodness of eggs, and large questions, like the safety of nuclear power. In fearful ignorance, they might even turn against science altogether, heedlessly throwing out the baby with the bathwater.

The social model that frames this sort of question is one of *rational choice*. It focuses on those points where a particular piece of knowledge might be expected to play an important part in people's lives—that is, when they have to make practical decisions to which this knowledge might seem relevant.

At first sight, this model might seem just a gloss on the deficiency model. Indeed, some utilitarian criterion like this is essential if we are to select items for public dissemination out of the practically infinite body of scientific knowledge. The measure of progress in public understanding of science would then be the extent to which people had already been given a sufficient understanding of the science that they might need on later occasions—or at least, the extent to which they might be prepared for such occasions by having a good general idea of the nature of science and of the scientific world picture and thus be capable of getting hold of the relevant scientific understanding as and when it was required.

The difficulty with this interpretation can be judged from some practical dilemmas drawn from our case studies:

- Should a Cumbrian farmer follow the advice of the agricultural scientists about grazing his sheep on pastures contaminated by Chernobyl fallout when it becomes clear that these scientists don't understand at all the familiar "science" of making a living out of a hill farm?

- What prior knowledge would really be of use to a person afflicted with a dangerous disease who must choose between two alternative courses of treatment?

- Where would you turn for further information about suspected environmental pollution—to the public library, to the county council, or to the firm that seems responsible for the danger, but that must surely know much more about it than anyone else in the neighborhood?

You can see that these are everyday questions that cannot be addressed properly, let alone answered, in terms of a rational choice model. Yet any assessment of progress in public understanding of science must surely include an account of how people—and society at large—can be helped to cope with such situations.

But the very language in which these questions are formulated may suggest a more calculated rationality than the events of life normally evoke. Ethnographers have begun to realize that analysis in terms of rational choice is not sustained when one looks carefully at the way that people talk about their motives and actions under such circumstances. This second model, also, does not generate a fully satisfactory frame for policy regarding public understanding of science.

THE CONTEXT MODEL

What seems to be missing is the *affective* component in social action. The question to be asked is, "What do people want to know in their particular circumstances?" In other words, a discussion of the full *context* of understanding has to be central to the analysis.

This discussion is obviously very complicated and will need to be carried much further before we can generalize reliably from it. But the research results we have been getting show very clearly that the theory and practice of public understanding of science should not underestimate the *incoherence, practical inadequacy, incredibility,* and *inconsistency* of formal scientific knowledge, as received and used by the public. Once again, this is not an attack on the scientific enterprise or its technological products: It is just a statement of the way in which this enterprise is actually presented to and perceived by people who are not themselves directly engaged in it.

There is not enough time here to present any of the research evidence for this disturbing statement,[2] but it may become more acceptable if I explain each of these points in a little more detail.

- *Incoherence:* People do not draw on stable, if fragmented and/or ill-conceived, *models* of the world, along the lines of textbook accounts of scientific knowledge. The little they retain of what they were taught at school is overlain and supplemented by the diverse representations of science that they encounter in the media and in many other aspects of life. What they pick up is not simply a filtered version of formal scientific knowledge: They have something that has meaning actively constructed by the processes and circumstances under which it is communicated and received.

- *Inadequacy:* The use that people make of formal knowledge in any particular situation depends on their needs of the moment and represents only one element in a complex and varied response. Not only do people rely heavily on the tacit, uncodified, but highly expert and rational knowledge that is shared in most work communities; they also engage with, select, or construct the scientific elements according to their own interests, involvement, personal and social histories, and other circumstances.

- *Incredibility:* People do not accept passively the knowledge presented to them by scientific *experts.* The credibility of a source depends strongly on its perceived interests in a particular context. This applies to individual scientists, scientific institutions, public bodies, and private enterprises.

- *Inconsistency:* Personal conflicts on social issues between scientific experts inevitably downgrade the privileged position of scientific knowledge. But public and private discussion helps people combine their scientific knowledge, ethical views, and tacit understanding of life into personal positions on controversial matters. In effect, people resolve the contradictions that arise by incorporating items of formal science into the whole knowledge

complex and making those contradictions *disappear* as such.

These, in outline, are the principles that seem to govern the way that people receive and use scientific knowledge. They have emerged from closely observed studies of people in particular situations—at work, in school, in the doctor's consulting room, or on their own doorsteps. As Beethoven said of good music, these principles are both surprising and expected: They are surprising if you come to this subject straight out of the lecture hall and research laboratory; they are expected if you reflect on your own experience of everyday life, thought, and talk.

To show the practical value of the context model as a correction to the deficiency and rational choice models, let me refer to one of the findings of the survey work carried out by John Durant and Glynis Breakwell.[3] This work showed a positive correlation between knowing about science and being interested in it. Despite all the conjectures to the contrary, the great majority of respondents reported themselves as "very interested" or "moderately interested" in news about discoveries and inventions—much more interested, in fact, than in news about sports, politics, or films. The paradox is that they also report themselves as worse *informed* about science and technology than about sports or politics. In other words, this general "interest" in the abstract, is likely to be much less influential than specific material "interests" in getting adults to understand science better.

This research also indicates that public understanding of science might be improved by taking more account of the social context in which understanding of science is acquired or exercised. The survey work showed, for example, that both attitudes toward science and involvement in scientific activities are significantly correlated with parental encouragement—particularly with parental attitudes toward science—irrespective of the age, sex, or social status of the respondent. Attitudes of friends toward school, youth culture, and achievement also correlate with variation in both attitudes toward science and the performance of science-related activities.

In attempting to measure progress in public understanding of science, we must always remember that scientific knowledge is not received impersonally as the product of disembodied expertise, but comes as part of life, amongst real people, with real interests, in a real world.

Notes

1. Royal Society of London, *The Public Understanding of Science*.

2. For some results from the studies coordinated by the Science Policy Support Group, see John Ziman, "Public Understanding of Science," *Science, Technology and Human Values* 16, no. 1 (Winter 1991):99-105; Roger Silverstone, "Communicating Science to the Public," *Science, Technology and Human Values* 16, no. 1 (Winter 1991):106-110; and Brian Wynne, "Knowledges in Context," *Science, Technology and Human Values* 16, no. 1 (Winter 1991):111-121.

3. John Durant, Geoffrey Evans, and Geoffrey Thomas, "The Public Understanding of Science," *Nature*, July 6, 1989, 340: 11-14; and Glynis Breakwell and S. Beardsell, "Parental and Peer Influences upon Science Attitudes and Activities," *Public Understanding of Science* (submitted).

John Ziman was professor of theoretical physics at Bristol University from 1964 to 1982. Since then he has worked on various scholarly and political aspects of the social relations of science. He was chairman of the Council for Science and Society for about 15 years, and has been director of the Science Policy Support Group, London, since it was founded in 1986.

Reprinted with permission from Bruce Lewenstein, ed., When Science Meets the Public, American Association for the Advancement of Science, 1992.

CASE STUDIES

In a case study of the Kansas City Museum, David Ucko draws on his experience transforming an older, eclectic museum into a science center, focusing on the need for a well-crafted mission statement. Cynthia Yao describes how people and resources coalesced around the notion of establishing the Ann Arbor Hands-On Museum, chronicling strategic steps in the institution's growth.

MISSION NOT IMPOSSIBLE *David A. Ucko*

Creating a mission statement appears to be a simple task. After all, don't all science centers have the same mission? Although science center missions may share similarities, your mission should be based on the unique strengths of your science center and the needs of the community you serve.

The purpose of most nonprofit organizations is to change people's lives. Your mission statement indicates what change you desire in your audience, and therefore what public service your science center provides. Thus the mission statement summarizes your basic purpose, your reason for existence. Commitment to this mission reflects opportunities or needs within your community, which must be matched by the capability of your institution to address them. The mission statement identifies your niche, and serves as a guide for your staff, volunteers, and board. It provides the long-range goals toward which day-to-day actions must be directed. The mission helps science center management target the use of limited resources of people and dollars.

The mission statement also lets those outside the organization, such as donors and potential supporters, know what your institution is striving to accomplish. Your success, and consequently, your ability to raise funds, will be measured by how effectively you carry out this mission. For example, the General Operating Support grant program of the Institute of Museum Services (IMS) uses this criterion as the basis of the review process to decide which institutions are funded.

It is easy to create a very broad, inclusive mission statement. However, such a statement can get you into trouble when political or financial interests push the institution in particular directions. Individuals or corporations may propose projects to the science center and even provide the funds with which to carry them out. Their ideas may be sound, but they must fit within your mission. Without a clear focus, it is too easy to become distracted and siphon away valuable internal resources from key programs. Sometimes, it is possible to redirect such projects in ways that directly support the mission. A well-defined mission acts as a valuable filter when you consider what exhibits and programs to offer.

During difficult economic times, science centers may need to cut back the programs they deliver, and may be forced to reduce the staff as well. A clear, agreed-upon mission statement becomes a tool that helps management make tough decisions about which services are most important to the organization and therefore must be maintained, while others are reduced or eliminated. It helps deal with science center programs that have become ends in themselves, rather than the means for achieving long-range institutional goals.

Finally, a mission statement is not a document developed as an assignment and then filed away to be forgotten. It must act as an ongoing guide for both short-term operations and long-term planning. When it no longer serves that function, it should be rewritten to make it serve that need.

CASE STUDY: KANSAS CITY MUSEUM

Prior to 1990, when I arrived at the Kansas City Museum, the institution viewed itself as a traditional museum. As stated in the fundraising case statement: "The first function of any museum is to collect, and it is the predominant reason for our existence." Since its incorporation in 1939, the museum had collected 250,000 objects in diverse areas, creating outstanding collections of costumes, native American artifacts, carriages, natural history specimens, and many other categories. Unfortunately, the museum lacked a clear focus. Our institution encompassed nearly everything but fine arts, which were the province of an older local museum.

Your mission is the basis for everything else that your science center does and will do.

We decided to re-examine the mission to clarify the museum's direction and to respond to changing community needs. When the museum was incorporated, the original emphasis stressed "history, natural science and industrial science." In fact, the museum hosted the region's first science fair in 1952, and became a member of ASTC in the late 1970s. During the recent past, however, regional history had been the museum's primary focus.

Although other local institutions collected and exhibited regional history, the Kansas City metropolitan area had no science museum. In light of this and the ever- increasing impact on the lives of Kansas Citians by science and technology, the board determined that the museum place increased emphasis on these areas. We wrote a revised mission statement to embody that thrust, creating a primary focus on science and a secondary focus on regional history. Still an important part of the museum, history now became a tool for understanding the past and present as a means of preparing for a future based on science and technology.

The museum's mission statement now reads:
> The Kansas City Museum promotes informal learning about science, technology, and history through participatory exhibits and programs that are both educational and entertaining. The primary audiences served by the Kansas City Museum are families and school groups from the region, as well as tourists.

The following is an elaboration of that mission statement:
> The Museum's primary *educational* mission is to stimulate public interest in science and technology, and to increase awareness of the influence that science and technology have on people's lives, and on the history of our region and its future. The goal of this effort is to enhance scientific literacy, and to encourage the pursuit of careers in science and technology.

> The Museum's *historical* mission is to preserve the region's heritage by collecting significant artifacts, specimens, and documents. These collections shall be used to support a secondary educational mission of placing the lives of today's Kansas Citians in historical perspective.

The Museum's *social* mission is two-fold: to reach children—with the intent to spark an interest in science at an early age, and to reach adults—especially those not predisposed toward science. Special target groups are women and minorities, groups that have been traditionally underrepresented in the science and engineering work force.

The Museum carries out its mission by offering experiences that are educational, entertaining, and actively involve its audiences. These experiences include exhibits and programs based on science and technology and the history and future of the Kansas City region. The Museum seeks to collaborate on its programs with other organizations whose goals support those of the Kansas City Museum.

To better carry out this mission, we now are in the process of creating "Science City," a separate new facility for the Kansas City Museum that will serve as an innovative stand-alone science center for the community. At the same time, we will renovate our existing facility, Corinthian Hall, as a stately home and history museum.

In summary, your mission is the basis for everything else that your science center does and will do. If you wish to create a firm foundation for your institution, prepare your mission statement thoughtfully, and then review it regularly. Although it is a difficult task, it's certainly not Mission Impossible!

Before becoming president of the Kansas City Museum in 1990, David Ucko was deputy director of the California Museum of Science and Industry in Los Angeles, and vice president for programs at Chicago's Museum of Science and Industry. Dr. Ucko was a chemistry professor before entering the museum field in 1979, and holds a Ph.D. in chemistry from M.I.T.

THE ANN ARBOR HANDS-ON MUSEUM: STEPS ALONG THE WAY

Cynthia Yao

The idea of starting a museum came to me in 1978 while I was completing my graduate degree in museum practice at the University of Michigan, after having stayed home with growing children for 12 years. I had two wonderful opportunities for travel to study museum programs in depth. One, through a $500 grant given as part of a summer minority museum internship from the Detroit Institute of Arts, allowed me to visit the education departments of 13 art and science museums on the East Coast. I also visited science and children's museums on the West Coast with a stipend from the National Endowment for the Arts. These visits, and all of the museum trips I ever made, were the seeds of inspiration for starting a participatory museum in Ann Arbor.

I began to organize meetings at several homes in Ann Arbor to discuss the concept and how to realize it. Early supporters met often and struggled with many issues, and made some decisions. We decided we didn't want the promotional type of exhibit or the push-button type of exhibit common in some science museums. We decided to size exhibits physically and intellectually to children about 10 years old. Thus our museum would not primarily serve preschool children (although many do love it). And we wanted to create exhibits in which the essential phenomenon could be examined by the participant, not hidden in a black box. We developed a set of general principles for exhibit selection and design (which we did not adhere to rigidly) to support our mission of bringing hands-on learning to our audience.

Excitement, anticipation, frustration, hope, disappointment, joy, creativity, satisfaction — emotions ran the gamut.

Another issue we debated early on was funding. Should we seek a large endowment of government funding to pay for a large staff, or have a modest operating budget that could be paid primarily from a modest admission fee? The more modest approach we chose meant that our exhibits would be built one at a time. We realized that funds for the building and its renovations would have to come from external sources.

CHOOSING A SITE

A critical issue discussed at early meetings was the site. Some people wanted a site away from downtown, with a big parking lot. We decided instead to create a museum closely integrated into the community.

Fate played a hand in this decision. The old fire station was empty for some three years and a handsome "state of the art"

new one was built next door. An article appeared in the *Ann Arbor News* stating that the city was looking for ideas for a new use of this beloved landmark. I had passed the building daily for years and now started to daydream about the idea, although I had not even been inside. I finally went to take a peek through the windows of the closed building and thought: "Wow! It's perfect! With its open spaces, very few changes would need to be made." Then another article appeared, stating that the local civic theater had approached the mayor with a plan to relocate in the fire house and he seemed receptive. I thought, "Darn it! That's also a great idea!" My husband suggested I write a letter to city hall, as there was a deadline for proposals. Little did I know that the city meant *financial* proposals for renovating the building.

I proposed sharing the facility. The idea of combining a museum and a theater was given front-page coverage: "Kids versus Culture"! So the city forged a partnership as a political solution. But later, the theater decided to pull out. Surprised by this, the city proceeded to impose stiff restrictions on us: We had to prove that a museum was feasible, become incorporated as a nonprofit organization, raise the funds to renovate the building, and report regularly to them on our progress. This was shocking—all I wanted to do was present an idea. I rallied neighbors and friends to come to city hall, and a groundswell of supporters began to meet weekly. We became incorporated as a nonprofit organization, and found a lawyer who volunteered to apply for 501(c)(3) tax-exempt status for us. Our monthly report to the city became our first newsletter.

LEARNING THE ROPES

What followed was four-and-a-half years of *Sturm und Drang*. Excitement, anticipation, frustration, hope, disappointment, joy, creativity, satisfaction—emotions ran the gamut. The group of museum supporters grew erratically at first. The publicity brought in many people, most of whom were strangers, and we had to learn to work together; some left the group when things didn't move as fast as they'd hoped. Meetings were scheduled at people's homes, where we literally passed a cup around to collect petty cash to pay for costs such as copying and mailing. It was difficult for me to play a leadership role when I had no experience in knowing what steps needed to be taken. We contacted other museums, the American Association of Museums, and ASTC, and read the few booklets and articles we could find, including information and advice gathered from my trips and visits to other museums. All this helped to give us direction. We were encouraged by the community's warm reception of our idea.

Most of us were educators who were totally innocent about fundraising. We needed and received professional, pro bono assistance from many museum people, lawyers, architects, fundraisers, exhibit builders, and designers.

Our scientific experts included three physicists, a mathematician, a psychologist, and several educators, plus several engineers and pediatricians. Everyone pitched in, bringing back slides from visits to other museums, making contacts, making copies of helpful literature, and holding many discussions which defined the exhibits and program focus.

Finally, after a year, we had achieved several major goals. The group's make-up changed to incorporate many community leaders who were known as "doers." Our first Community Advisory Council was formed, which later evolved into our first Board of Trustees. This was important because it broadened our base and we gathered more experienced, accomplished people who had the ability to help us raise funds. We were successful with our first grant proposal, a National

Register Grant-in-Aid, for nationally registered historic buildings like ours. It was a two-to-one match whereby we would receive $60,000, which would yield $180,000 when matched, close to the estimated $250,000 that we needed for renovation.

Three of us went to the Exploratorium—a physicist, and a mathematician, and me—to learn about their operations, especially their exhibits. We created a slide show from other museums, showing what was done in other communities, and we started our fundraising campaign with help from University Microfilms International, then a subsidiary of Xerox Corporation, which adopted us as a community relations project. They covered all our printing costs and helped us with marketing.

CREATING OUR FIRST EXHIBITS AND RAISING FUNDS

By the next year, we had received a small grant from the University of Michigan to create a prototype traveling exhibition to demonstrate the "hands-on" concept to the community. We created our first seven exhibits, adapting six exhibits from the *Exploratorium Cookbook* which were favorites when we were interns. We did not want to be creative at that time—we wanted proven, successful exhibits. We moved our newly built exhibits around campus and downtown in any free space we could find. We gathered more than 50 volunteers to staff the exhibits and created a manual of instructions. This was our first taste of museum programming and we loved it. We found talents to build exhibits within the community and through the University's Marine Engineering Department.

After moving these exhibits seven times within three months, we found a temporary summer home for them at the University's campus. This time we staffed the exhibits with summer CETA students. (CETA is a now obsolete federal government program that paid salaries.)

That fall, the Ann Arbor Public Schools took over the exhibits, created a slide show, hired a CETA teacher, and moved the exhibits to 26 schools within the district, where they were seen by more than 15,000 students. Later, three other outlying school districts took the exhibits and an estimated 40,000 more students were reached. The exhibits traveled for more than two years, making friends for us, while we were raising the renovation money we needed.

With help from the pros and much trial and error, we learned how to raise money. We learned not to go looking for funds until all our plans and people resources were in place. When we could tell a convincing story, complete with prospectus, people listened. We learned to persevere; if Plan A didn't work, we went to Plan B. And so we made it.

MANY "FIRSTS" AND TENTATIVE STEPS

The museum opened, 10,000 square feet on the first two floors of the fire house, with limited hours, on October 13, 1982, staffed by one overworked director and several volunteers. Within three months, a part-time assistant director, who had been a volunteer for several years, was hired. The museum was an immediate hit with only a $50,000 budget, which we earned as we operated.

During the first day, about 10 of the 25 exhibits broke down, a real test of our exhibit design. Luckily, they were things we could fix immediately. Dick Crane, our chief physicist and designer, could not leave town. Staff had to learn to be handy in a hurry. It wasn't just exhibits—we also had to keep the place clean and inviting. Our exhibit repair policy became "never repair something the same way twice—improve it!"

We fleshed out our themes further, improved exhibits, improved graphics, repainted, added new exhibits, and expanded our Discovery Room. We began filling two floors of galleries with exhibits. We tried out education programs, including Saturday classes, preschool classes, Kitchen Chemistry, Chocolate Chip Geology, and summer classes. We added special events, such as "Bubbles and Balloons" and an "Open Your Heart" benefit. We had more scheduled groups, and we added weekend demonstrations, 15-20 minute shows on weekends with the theme changing monthly.

Each staff member performed multiple functions and even changed jobs willingly after a year. We had to learn how to operate the cash register, do accounting (the first four years we had a volunteer in this position), and run a gift shop. We created a graphically pleasing newsletter and introduced computers into our business operations.

Our museum generated funds through admissions, memberships, gift shop, and small benefits. We beefed up each program, and success escalated annually.

But this was also a delicate and difficult time. Three volunteers moved into salaried positions. We learned to separate board members, program committee, and advisory people from staff regarding who makes decisions. The staff worked exceedingly hard, but were excited by the daily adventures. We willingly put all our energy into the museum and got spouses, families, and friends involved. We had estimated 6,000 to 10,000 visitors, and instead we got 25,000 the first year, which had increased to 56,000 three years later.

Success forced growth. We hurtled along, swept by the need to meet public demand. For example, we needed an elevator to provide access to the second floor for people with disabilities, which triggered planning for another capital campaign.

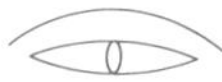

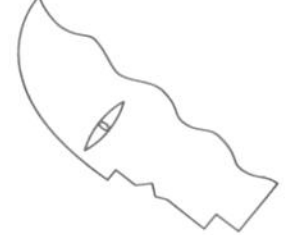

THE SECOND FOUR YEARS

Our second capital campaign, begun and completed in 1985/
86, was more comprehensive and easier to do. With a track
record and some hard-won savvy, we raised $400,000,
including a Kresge grant and a large state grant. A second
renovation was begun and completed. The museum was
closed for five months. We added new staff, reorganized, and
made improvements. We added 50 percent more space,
especially on the third and fourth floors of the building, and
many of the amenities that we hadn't thought about in the first
renovation, including more bathrooms, an elevator, carpeting,
a volunteer room and kitchen, a library, enlarged class storage
space, and exhibits supply space.

We added new exhibit themes with the new floors—Math
Games and Puzzles, Computers, and How Things Work; and
we consolidated other light and optics exhibits onto the new
third floor. We also made some internal operational improvements:

- offices on each floor to maintain staff contact with the
  public

- an annual audit

- staff rotation of responsibility of floor manager for the day
  and weekends

- paid, full-time gift shop manager

- more summer classes, more classroom space outside of
  building

- improved handling of groups, procedures for going up
  and down stairs

- improved volunteer training/recognition; expansion of
  volunteer program

- use of more work-study student and credit-study student
  programs

- computerized offices

We soon had eight full-time employees, 20 work-study
students, and 200-400 volunteers. We became able to offer
better staff benefits—medical insurance and pensions, and
professional development opportunities, like going to conferences and workshops. We have received five General Operating Support awards from the Institute of Museum Services, and
through our two major capital campaigns and successful
grants, we have raised closed to $750,000 in the past nine
years.

We collaborated with other Michigan museums to get state
funds. We also began an outreach program to reach
underserved audiences. Then we were awarded a grant from
the National Science Foundation for outreach, to develop
traveling exhibits to circulate nationwide.

The museum is now a participatory museum of more than 190
interactive exhibits that focus primarily on science, with some
art and culture-related themes. In 1986, it was cited as the
"best of the best" places to take children to on a "junket of
joy" trip, out of 90 places visited in Michigan by the *Detroit
Free Press*. In the summer of 1988, ASTC held its first Institute
for New Science Centers at the University of Michigan and
used our place as a model start-up museum. Our attendance is
now more than 100,000 annually, and we have an annual
budget of $500,000. After only four years, we are bursting at
the seams again.

FOR THE FUTURE

We are now discussing the possibility of holding a $1-1.5

million campaign. We need to expand to double our current size of 15,000 square feet, possibly into the building next door or one down the street. We need more classrooms, offices, workshops, storage space, and an auditorium for 100-150 people. We are planning a new focus on preschool and high school age groups. We also would like to have some flexible exhibit space, with the capacity to handle large traveling exhibits, about 3,000 to 4,000 square feet.

Can we do it again? Sure, if we stick to our proven formula for success: Learn from other institutions, but do your own thing.

Keep excellence as a goal, and remember that visitors and education are your most important assets.

Cynthia Yao is executive director and founder of the Ann Arbor Hands-On Museum in Ann Arbor, Michigan. She has a B.A. in fine arts from Emmanuel College, Boston, and a M.M.P. (Master of Museum Practice) from the University of Michigan. Ms. Yao serves on the board of directors for the Ann Arbor Area Chamber of Commerce, and is a member of the Statewide Board Committee of Governor's Quest Program.

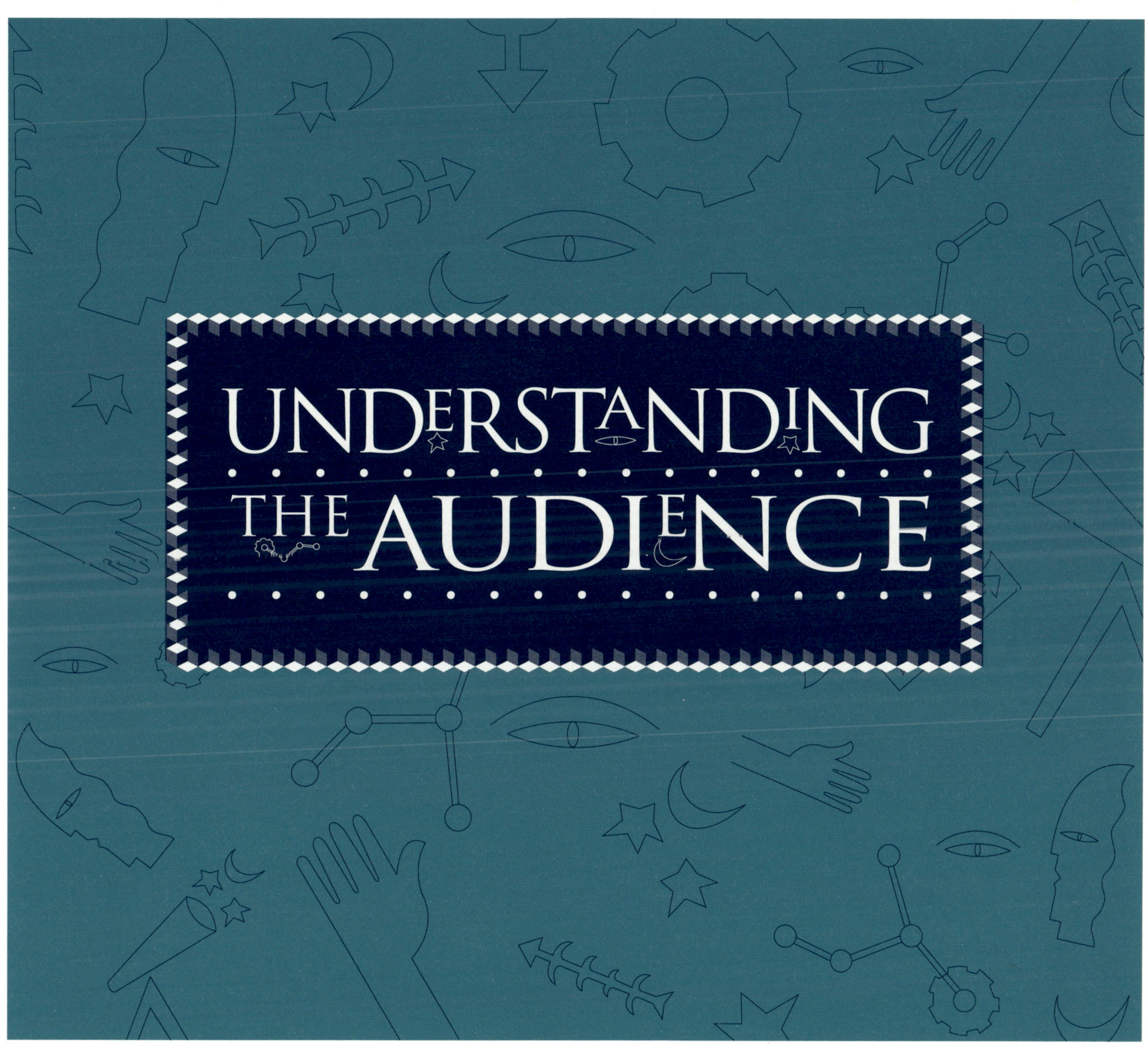

UNDERSTANDING
THE AUDIENCE

When you open the doors of your new science center, who is likely to come? What will they expect of you, and what can you expect of them? Here's what we know about the audience for science centers in general. Your own attendance will probably vary from these norms. Plan to survey your audience regularly, both to measure your impact and to plan for improvements and growth.[1]

HOW MANY PEOPLE COME TO SCIENCE CENTERS?

People planning science centers look at the attendance records of existing institutions and try to extrapolate how many people will visit their new place. Sometimes they examine the size of the metropolitan area, competing attractions, and marketing budgets to develop a sense of how large a clientele their community will support. A closer look at data from existing science centers and museums shows that this approach has little predictive value.

The size of the metropolitan area—and, to a lesser degree, the amount of marketing money spent—have little bearing on visitation. People come to science centers, it turns out, because of what is offered, and how much of it there is.

In a first-rate multivariate statistical analysis of data from 38 ASTC members performed in 1989 for the National Science Center in Fort Gordon, Georgia, attendance was shown to correlate most closely with exhibit space.[2] The more exhibits there are to see, the more people come to see them, regardless of the size of the community. Everyday experience bears out this notion. Think of the American Museum of Natural History and the Manhattan Children's Museum, just a few city blocks apart in the same demographic context, but orders of magnitude apart in size of exhibit space—and in numbers of visitors.

But while attendance parallels physical size, it does so only roughly. Clearly, the quality of the experience offered comes into play. We don't know precisely how people select the frequency of their visits, but we do know that they are most influenced by word-of-mouth. If a trusted source (family or friend) says, "that place is terrific," they are likely to go see for themselves. If you build a science center that is "good," you can be sure that people will come.

For purposes of planning, we can calculate from ASTC survey data that every 10,000 square feet of exhibit space generates an average of 100,000 annual visitors. In actual practice, there is considerable variety around this number—this is not a predictive model. Note that this measure neglects the effects of programs, theaters and planetaria, and collections (if any) and so does not accurately represent the thrust and scope of an institution. Nor does it describe how crowded or inviting the exhibit floor feels, or how visitation varies over time. But it does provide an empirical guideline for sound fiscal planning: If you build and fill 15,000 square feet of exhibit space, you can prudently count on receiving 150,000 visitors and the admissions revenues they bring (at the 1986 rate of visitation).

People come to science centers because of what is offered, and how much of it there is.

Many science centers include large-screen format theaters and other special attractions with their own distinct visitation patterns. A number of planners have developed rules of thumb for calculating visitation for such attractions.[3] Sometimes science center staff consider these features to be a means for drawing new audiences (for example, young adults who want night-time entertainment) and revenues. Staff hope that once inside the institution, these special purpose visitors will utilize the rest of the center. In fact, the first part of the hypothesis appears to be true: Different audiences do come. But they don't necessarily patronize the rest of the facility, having satisfied their own goals for venturing forth. Science centers that wish to add non-exhibit attractions, such as Imax theaters, must carefully evaluate the impact on and fit with the center's mission; the commitment of time, money, and effort may or may not serve the center's larger purposes.

WHY DO PEOPLE COME?

People come to science centers to have a good time in a wholesome setting where they can bring friends and relatives, exercise their curiosity, and learn some science. We don't have field-wide data that would allow analysis of which of these attributes of a science center visit is most important to different categories of visitors. A few studies at different institutions, taken together, suggest that, within the social context of the excursion, both having fun and learning science are important to most people.[4]

Except for organized groups like school field trips, visitors come voluntarily, during their leisure hours. In effect, science centers compete with other leisure-time pursuits for visitors' time and money. Some people in the field take this to mean that science centers should adopt characteristics of the entertainment industry: They should market to visitor preferences, dazzle with novel technology, and provide facilities to accommodate large crowds. Given the success theme parks have had, some science center staff take very seriously Disney's carefully honed methods for attracting and accommodating large numbers of people.

But other characteristics of theme parks are incompatible with the mission of science centers. Parks provide seamless experiences in which the sometimes passive visitor makes no mental effort, and technical wonders go unexplained, inaccessible to questioning. Science centers expect their visitors to work harder than theme park visitors do; visitors are expected to take an active role in exploring the facility, asking questions, and seeking explanations during much of their visit. Ideally, this work is its own reward, because well-constructed exhibits and well-structured programs help visitors gain new insight, itself a source of great satisfaction.

To the extent that science centers facilitate this kind of learning, they meet their visitors' expectations, as well as their community's needs. So far, science centers have enjoyed the good will of their visitors. Not only do people come voluntarily, they pay attention willingly, even those in organized groups, to what is offered by a responsible institution. Visitors trust the information and experiences science centers provide, perhaps as an extension of the affection with which they have regarded older, natural history museums. As long as the visitors' attention is appropriately rewarded—with experiences that combine leisure and learning—science centers can keep the public's good will. They can maintain their privileged position in the community and distinguish themselves from commercial ventures, to visitors and funders alike.

WHO COMES TO SCIENCE CENTERS AND WHAT DO THEY DO?

Half of the science center audience is at least 18 years old, and half is younger.[5] The dominant users are family groups, who come on weekends, holidays, and summers; science centers are generally more crowded at those times. The single most crowded day is likely to be a rainy Sunday in July or August (except that, in the U.S., the day after Thanksgiving breaks all records). Fewer visitors come early or late in the day, so family visitor density is greatest mid-afternoon.

School groups come on weekdays (in times families don't), from October in increasing numbers through early June (in the U.S.), and day camp groups come in summer. School groups come when local buses are available; often this means from 10:00 am to 1:00 pm. In general, total attendance dips in September, when families and schools are reorganizing after the vacation. Special exhibitions, particularly of animated dinosaurs, can skew attendance patterns markedly.

Children do retain a good bit of the science they encounter during the visit. Having fun plays an important role in learning.

A typical family visit lasts from one-and-a-half to two-and-a-half hours. Some science center staff believe that a lunch facility helps families prolong their visits. Having a place to sit and find refreshment certainly makes a family's visit more comfortable, if not lengthy.

When families with children tour the center, they generally stick together and talk about the displays, helping each other to understand and value what is going on. For other groups, the amount and kind of social interaction appears to depend on the group's composition. (One study showed that a male/female pair of visitors are unlikely to converse, but very likely to read the text![6])

School groups from first grade through high school utilize science centers, with the preponderance from elementary school. Because children have different teachers for different subjects in middle and high school, excursions are much harder to arrange. Typically, a visit lasts one-and-a-half hours and may include a special presentation or sit-down workshop arranged in advance. Generally children are also invited to roam the exhibit halls, under the relaxed supervision of teachers and parent chaperones. Almost always, when schoolchildren first descend on the exhibit floor, they dash from corner to corner, twisting knobs and tweaking levers without pausing to observe results, excitedly checking it all out. After about 20 minutes, they settle down and engage earnestly with the displays that catch their (or their friends') attention.

Critics grant that schoolchildren have fun during science center visits but question whether they learn anything.[7] For critics accustomed to the controlled quiet of the academy, behavior at a science center can be distressing. But while the high level of activity children display at science centers does not promote systematic thinking, it does feed curiosity and encourage personal choice, both intrinsic motivators for learning. There is some evidence, moreover, that children do retain a good bit of the science they encounter during the visit.[8] Having fun plays an important role in learning: In order to learn about something, you have to get close to it, and if getting close to something is enjoyable, you are more likely to do it.[9] This applies equally to school groups and to general audiences.

Almost all science centers host school groups and over half served more than 25,000 school children in 1986.[10] On average, science centers receive 24 percent of their visitors from the schools, but the actual proportion varies greatly with the size of the institution. Older, smaller centers tend to serve a higher proportion of school groups; large centers serve a higher proportion of general visitors. Perhaps this reflects differences in mission between large and small centers. Or perhaps it reflects some kind of structural constraint that limits the number of schoolchildren a large museum can handle.

(Some people say the size of the lunchroom restricts the number of classes that can be comfortably accommodated within the busing schedule.) New museums tend to concentrate first on the life of the exhibit floor, developing resources and methods for serving the school community later, as they mature.

WHAT DOES THE AUDIENCE NEED?

Several streams of research—at museums, about science communications, and in various branches of psychology and learning theory—converge on a single notion: Members of the audience play an active role in creating their own experience at the science center. Visitors' prior knowledge (accurate or not), personal interests, social interactions at the center, and preferred learning styles all affect what they see and do. The more science centers understand and accommodate visitors' thinking, the more, and happier, communication they can accomplish. But accommodation is a tall order, given the wide range of ages, backgrounds, technical competences, and social groupings found among science center visitors. How do we know what different people need? And how can we possibly serve all of them, given limited space and resources?

Some science center staff simplify the task by focusing their communications strategies on children, assuming adults will act primarily as chaperones. Thus they feed a widespread perception that science centers are for kids. But this gives half their audience short shrift. Science centers can and should be designed for adult visitors *and* for children, keeping their different capabilities in mind, as well as the probability that they will interact with one another at least sometime during their visit. Nonetheless, it may be true that science centers have the greatest impact on the young.

With thoughtful planning and design, you can appeal to and intrigue both young and old, expert and naive. First, you try to select phenomena or experiments that are intrinsically interesting or beautiful, so that everyone wants to examine them—like a misty tornado forming right before your eyes. You can provide ways to explore through sight, touch, sound, and motion, including at least one obvious thing that a grade-school child can do unaided. Most adults and most children will dig right in, although with different degrees of inhibition and activity.

You can also provide "layers of interpretation" that different visitors can use to dig deeper, at will. There may be additional apparatus for altering a variable (rotational speed, for example) and seeing the results. Or there may be additional text that relates to local weather patterns or tornado mythology (addressed to adults, who may reinterpret some of it to their kids). Or additional graphic and textual materials on geology or meteorology might be available via computer or interactive videodisc nearby. There might even be a person to whom questions could be directed.

> **Members of the audience play an active role in creating their own experience at the science center.**

Not all visitors will use all these interpretive layers —nor should you expect them to. Rather, you can expect that in the course of their visit, people will find and engage with material that corresponds to their own interests and preferences. If you stand next to one particular graphic panel or a videodisc display, you will observe that the majority of visitors ignore it. But if you follow (adult) visitors over the course of the entire

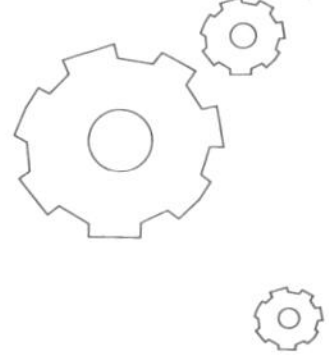

visit, you will see that they stop and linger at a few places and read most of the graphic panel there[11] or spend as much as 10 minutes at a videodisc display of their choice.

Another strategy science center staff use to accommodate varied audiences is to divide up the space. Sometimes they mount temporary or traveling exhibitions on technical or policy issues primarily of interest to specialist or adult audiences. Sometimes they set aside space for preschool children who can't keep up with older siblings (parents are usually asked to supervise the toddler while the older child roams free). Sometimes space and personnel are devoted to a program or instrument that serves only one audience intensively—middle school teachers, for example. How much subdividing a science center does depends on its mission—on what it wishes to accomplish.

Studies of visitor behavior at museums point also to visitors' need for physical orientation and comfort. At a minimum, visitors need to understand how to get into and around the facility without getting lost. This translates into directional signage (starting on the approaching streets), and places to rest and reconnoiter (either in the exhibit areas or in addition to them). School visitors need enough accessible bathrooms so children can be accommodated readily. Visitors get frustrated when they are funneled down blind alleys, and they need room for group interaction around and between displays. Accommodations for people with disabilities, required by law, are most easily built into facilities from the start.

Visitors also need to understand content options so that they can exercise choice in structuring their visit. Traditionally, museums have used many techniques to orient visitors to the museum's content: films, information desks, maps, computers. While there's no preferred way to go about providing it,

orientation to content, as well as to the building, contributes to a satisfying visit.

WHO DOESN'T COME TO SCIENCE CENTERS?

Two surveys, one in the U.S. and one in Britain,[12] show that most adult visitors to science centers come from the middle and upper classes. Based on the observations of other European and Asian science center staff, we can assume that the pattern holds true practically everywhere. In every country, then, science center staff grow accustomed to the habits and preferences of their ready audience, and plan with them in mind. Programming, marketing, funding, and even staff recruitment perpetuate the middle class pattern.

Science centers that want to diversify their audiences must make a self-conscious, concerted effort to attract people who normally do not come.

Survey researcher Jon Miller's studies show that in the U.S., attentiveness to science—including going to science centers—is limited to 20 percent of the adult population.[13] Miller's work on the "attentive public" has been widely discussed by people involved in science communication. What are its implications for science centers' efforts to reach an increasingly broad audience?

Miller divides the public into a four-tiered pyramid regarding the formulation of social policy involving science and technology: A few thousand science policy-makers are at the top, followed by the "attentive public" to whom policy-makers normally address themselves. This 20 percent of the populace

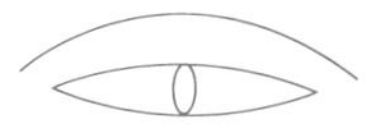

includes adults who, by Miller's definition, show 1) a high level of interest in science and technology, 2) factual knowledge, and 3) a regular habit of science information acquisition (visiting a science center is one indicator of this habit). Next Miller distinguishes the "potentially attentive"—about another 20 percent of the American adult public that is interested but not knowledgeable. The rest of the public, some 60 percent who are neither interested nor informed, forms the pyramid's base.

Miller and others argue that communication about science is effective with the attentive tier, and may serve to enlist the potentially attentive, but that it is generally lost on the inattentives. In the interests of efficiency, Miller argues that we should communicate science to those who absorb it and are prepared to do something about it—i.e., the attentives. Even the attentives, by Miller's measures, are not generally "science literate." If they are to act effectively as ratifiers of science policy, they need more information and better understanding.

Miller's work has stimulated others to look more closely at the audiences for science communications. One recent study by the Public Agenda Foundation is of particular interest to science centers.[14] The study set out to test whether ordinary Americans could make consistent, well-formed policy decisions on complex scientific issues. A demographically representative sample of 400 citizens was shown a film about environmental issues followed by a 45-minute, even-handed discussion. Then, when polled, the citizens made essentially the same policy recommendations as a panel of 418 scientists who had been surveyed by mail. So, given an appropriate intervention, "inattentive" citizens do as well as anybody else.

Science centers could provide "appropriate interventions" on complex, science-based issues (similar to what was done by the Public Agenda Foundation) that would enable all members of the adult public to participate thoughtfully in science policy formulation. While Miller's argument to address the top half of the attention pyramid is convenient, it clearly does not define what is possible or desirable in the science center setting.

Science centers that want to diversify their audiences must make a self-conscious, concerted effort to attract people who normally do not come. People of lower socioeconomic status are less likely to have "the museum habit" or may not feel welcome at a place where they see few people like themselves. Some science centers have set aside time and resources to make contact with leaders of underrepresented constituencies and to include them in planning, and they have gradually succeeded in expanding their reach into new communities. Progress may be slow, but it is critical. It is often implicit, if not explicit, in a science center's mission, and it can justify governmental or philanthropic support.

Notes

1. For an introduction to visitor surveys, see chapters 2 and 3 in Ross Loomis, *Museum Visitor Evaluation*.

2. National Science Center for Communications and Electronics, *Estimates of Annual Number of Visitors to the NSCCE Discovery Center: Executive Report*.

3. See, for example, Peter Anderson, *Before the Blueprint: Science Center Buildings*.

4. See, for example, an early, comprehensive study by Michael B. Alt, "Four Years of Visitor Surveys at the British Museum (Natural History) 1976-79."

5. Susan McCormick, ed., *The ASTC Science Center Survey: Administration and Finance Report*.

6. Paulette McManus, "It's the Company You Keep...the Social Determination of Learning-related Behaviour in a Science Museum."

7. See, for example, Michael Shortland, "No Business Like Show Business."

8. Jeffry Gottfried, "Do Children Learn on School Field Trips?" See also Minda Borun et al, *Planets and Pulleys: Studies of School Visits to Science Museums.*

9. The role of intrinsic motivation in learning has been described by Mihaly Csikszentmihalyi. For an introduction to his work, see "Human Behavior and the Science Center."

10. Mark St. John and Sheila Grinell, *The ASTC Science Center Survey: An Independent Review of Findings.*

11. Minda Borun and Maryanne Miller, "To Label or Not to Label?"

12. The U.S. data is from Katherine Dimmock, "Models of Adult Participation in Informal Science Education." The British data is from an unpublished study by D. Cox, "Attitudes to Science Among the Public Visiting Science Centres/Exhibitions."

13. Jon Miller, *The American People and Science Policy,* especially chapter 4.

14. John Doble and Amy Richardson, "Scientific Issues and Thoughtful Public Involvement: A Case of the Impossible vs. the Inevitable?"

V I E W P O I N T

In a ground-breaking study, Marilyn Hood investigated leisure time preferences of both people who visited and people who didn't visit a metropolitan art museum. Some of her conclusions about the audience for cultural institutions apply to the science center setting.

STAYING AWAY: WHY PEOPLE CHOOSE NOT TO VISIT MUSEUMS

Marilyn G. Hood

Those of us who are museum professionals have frequently been puzzled by the elusive masses who never enter our museum doors—the nonparticipants. With all the treasures we offer, why don't we attract a broader spectrum of the public, a larger audience, a substantial clientele that comes regularly rather than just for blockbusters? Why do programs that are successful in one museum or with one group fail to garner equal response with other audiences?

Over the past half century we have tried numerous research techniques to gain answers to these questions. We have tracked visitors' traffic patterns, timed their stops at exhibits, observed them with time-lapse photography and interviewed them in the museum to find out what was satisfactory about their visit. It is not surprising that none of these *in*-museum studies has told us why the majority of the public does not visit museums.

From hundreds of such surveys in the United States and Canada, however, we have learned the demographic characteristics of those who do patronize museums: They are likely to be in the upper education, occupation, and income groups, younger than the population in general, and active in other community and leisure activities. Nevertheless, these demographic data have not indicated the *reasons* why some adults choose to frequent museums and why some do not, or why nonparticipants don't love museums.

It is apparent, then, that merely analyzing demographics will not reveal what these groups value in their leisure experiences. Instead we need to focus on how individuals make decisions about the use of their leisure time and energy, to concentrate on the *psychographic* characteristics of both current and potential visitors—their values, attitudes, perceptions, interests, expectations, satisfactions. Once these factors are identified, we can examine how nonparticipants differ from participants in order to determine whether or not museums are offering or can offer the kinds of experiences that nonparticipants value and expect. Then we can develop ways, within the scope of our organizations and our abilities, to reach these elusive audiences.

In carrying out such a plan, the basic step is recognizing that people make *choices* about how they will use their leisure time and energy. We often assume that because we regard museums as unique and valuable, the public will similarly cherish them and want to share in them. Individuals do not just naturally gravitate to museums or to any other leisure place, however, no matter how worthwhile or unique it may be. Instead, before making selections, they consider which of several competing alternatives appears to offer them the most rewards, the greatest satisfactions—and they make their choice based on what will satisfy their criteria of a desirable leisure experience.

What are these criteria by which individuals judge leisure experiences, including museum visits? A review of 60 years of literature in museum studies, leisure science, sociology, psychology, and consumer behavior identified six major attributes underlying adults' choices in their use of leisure time. They are, in alphabetical order:

- being with people, or social interaction

- doing something worthwhile

- feeling comfortable and at ease in one's surroundings

- having a challenge of new experiences

- having an opportunity to learn

- participating actively

Not every person values all of these attributes, and some are more pertinent to certain activities or places than to others. But all are fundamental criteria by which individuals make decisions about leisure.

To test how these criteria affect museum participation, a carefully designed and tightly controlled research project was undertaken in 1980-81 in cooperation with the Toledo Museum of Art, Toledo, Ohio. The purpose was to obtain quality information that would be useful for long-term decision making by the Toledo Museum and museums in general. To achieve valid and reliable results, several preparatory steps were taken: A 12-page questionnaire was developed, based on previous research and theory, and was tested and revised; the survey sponsor was identified as the Ohio State University, so respondents would not be biased in their answers to questions about museum going; a computer program generated telephone numbers for a probability sample of Toledo metropolitan area residents to be interviewed by telephone (in a probability sample, each person in the population has an equal chance of being selected, which assures that the sample is representative of the population); and 35 museum volunteers were trained to administer the 20-minute questionnaires.

The volunteers secured telephone interviews with 502 residents from across the Toledo metropolitan area (urban, suburban, exurban, rural) over a three-week period in spring 1980, and the data from the questionnaires were thoroughly analyzed by several sophisticated statistical tests, with the assistance of a statistician and a computer. A detailed report, including 67 tables, described the relationships between leisure attitudes and values and between museum going and demographic characteristics, and it outlined how these findings can be applied to museums generally and to other leisure-cultural organizations.[1]

The Toledo metropolitan area was a suitable locale for a major research project because its population was large enough to include representative socioeconomic, educational, and age groups so that a probability sample of telephone respondents would accurately reflect the opinions and values of the community and the study results be applicable to other locations. In addition, it was assumed that the prestigious Toledo Museum of Art was well enough established to be identifiable by all groups within the community. This assumption was substantiated, since all respondents in the sample, regardless of socioeconomic status or length of residence, knew of the museum, even if they had never been there. Also, the types of problems that the Toledo Museum faces are applicable to a greater or lesser extent to most museums, regardless of size, and the staff recognizes that it is in competition with other community activities for people's time, attention, and energy.

Comments by Telephone Survey Respondents

"The museum is a lovely place. But I like a noisier place."
– A motorcyclist who had visited the museum as a child.

"I went to an opening alone and there was no one to talk to."
– A young professional, new to the city.

"You have to be quiet there, like in a library or church."
– A young mother of small children.

"Museums are places the kids ought to see— like zoos."
– A father of young children.

Two major aims of the study were to determine how important the six leisure attributes were to the respondents and to ascertain their preferences for certain leisure activities and places. The study also assessed respondents' attitudes toward art museums and gauged their level of socialization toward 22 activities, including museum going. This article briefly discusses the results of the first two aims.

The results clearly show that our traditional assumptions about museum audiences are unfounded. Our long-held belief that there are just two audience segments—participants and nonparticipants—is incorrect.

There are *three* distinctly different audience segments in the current and potential museum clientele, based on their leisure values, interests, and expectations: frequent participants, occasional participants, and nonparticipants. Each group seeks different values and experiences through leisure activities, including museum going. Moreover, people decide to be or not to be involved in museums on the basis of how they evaluate the six leisure attributes and on how they were socialized—any family and other childhood influences— toward certain types of activities. Though other museum studies have identified levels of participation, they have not probed the *reasons* why audience segments develop and are maintained. Now, with these results, we are able to identify attendance patterns by leisure values and to show that leisure choices, although they may be correlated with demographics, are not determined by demographics.

This is strikingly clear when we examine the profiles of the three audience segments. The frequent visitors—those who go to museums at least three times a year (and some as often as 40 times a year) highly value all of the six leisure attributes and perceive all of them to be present in museums. The three that they value most are distinct for this group: having an opportunity to learn, having a challenge of new experiences, and doing something worthwhile in leisure time.

Though the frequent visitors constitute a minority of the community (14 percent in the Toledo metropolitan area), they account for 45 to 50 percent of museum visitation (the Toledo Museum's annual visitation is 300,000 to 400,000). These are

the people who are usually interviewed *in* the museum; hence, they fit the typical museum visitor demographic profile.

These loyalists go to museums wherever they are and whatever is showing, because some time ago they chose to place museums on their leisure agenda. Since their experience with museums has developed over time, they identify with museum values and understand the "museum code" of exhibits and objects. Museums are satisfying places for them because they find that the three leisure attributes they value most highly are regularly available in substantial quantities in museums.

For frequent attenders, the benefits offered by museum visits consistently outweigh the costs (time, money, travel, mental saturation, fatigue, inconvenience). Because they come so often, we museum professionals should make sure the museum is not a static place remaining always the same; these visitors want to find the challenge of new experiences on a continuing basis in their leisure activities.

At nearly the opposite pole from the frequent participants—in leisure values, preferences, and expectations, as well as most demographic characteristics—are the nonparticipants (who represented the largest segment, 46 percent, of the Toledo metropolitan community). In their leisure experiences they most value the three leisure attributes that were less important to the frequent visitors: being with people (social interaction), participating actively, and feeling comfortable and at ease in their surroundings. And, underscoring their differences, they rank low the three attributes the frequent visitors preferred.

Generally, nonparticipants as children were not socialized into museum going; in fact, they are likely to have adopted more cultural activities as adults than they were acquainted with as children. We museum professionals and devotees need to be wary, however, of labeling these persons as apathetic or uninvolved simply because they do not participate in cultural activities. Their interests and commitments lie elsewhere, and they choose leisure experiences that compete with museum going because they find more of what is satisfying to them in activities that emphasize active participation, casual and familiar surroundings, and interacting socially with other people.

There are three distinctly different audience segments in the current and potential museum clientele, based on their leisure values, interests, and expectations.

Nonparticipants perceive that these three leisure attributes— the ones they value most highly—are not present at all in museums, or are present in such small amounts that investing themselves in a museum experience brings minimal benefits. They perceive museums to be formal, formidable places, inaccessible to them because they usually have had little preparation to read the "museum code"—places that invoke restrictions on group social behavior and on active participation. Sports, picnicking, visiting, and browsing in shopping malls better meet their criteria of desirable leisure activities.

The most notable finding from this research involves the occasional participant—those who visit museums once or twice a year (40 percent of the Toledo metropolitan community). We have long assumed that museum visitors, regardless of frequency of attendance, share many common values, interests, and characteristics. The research results emphasize, however, that the occasional visitors are distinctly different from the frequent visitors in their socialization patterns and leisure values. In fact, they more closely resemble the nonparticipants.

The occasional visitors were socialized as children into activities that emphasized active participation, entertainment, and social interaction. As adults, they maintain high levels of participation in these types of activities—outdoor experiences such as camping, hiking, swimming, skiing, ice skating; playing a musical instrument, or engaging in arts and crafts; going to amusement parks and movies; sightseeing and attending sports events.

Family-centered activities are much more important to the occasional participants and nonparticipants than they are to the frequent participants, who are more likely to visit the museum alone. Places like parks, zoos, and picnic areas that are natural centers for family outings and for extended-family visiting attract the occasional participants because they offer all three of their most highly valued leisure attributes. Going to outdoor art and music festivals and participating as a family in a craft or discovery workshop also meet their criteria of desirable leisure experiences.

Occasional participants, who value comfortable surroundings in their leisure places, feel that museums offer little in the way of comfort—not simply physical comfort but a feeling of "this is where my friends and I belong, a place where I feel at ease and am able to cope with the message." For this group, leisure is equated with relaxation, which is more akin to interacting socially with a family or friendship group than it is to the intense involvement in a special interest that is evidenced by museum enthusiasts. Because these persons do not feel entirely at home in a museum setting, the presence of a support group—family, club, co-workers, friends—provides

social approval and validation on a visit.

Occasional participants perceive that some of the attributes they value in leisure experiences are available in museums, but not in sufficient quantity to warrant regular visits—especially when compared with the benefits afforded by competing interests. Consequently, they come for the special occasions, the major events, the family days, which seem to promise them greater fulfillment of their expectations and wants. Since museum values and intentions more closely resemble those of the frequent visitors, museums generally offer or emphasize the very qualities that are least appealing to the occasional participants and nonparticipants, who are looking for significantly different leisure satisfactions.

The findings from this study provide a new perspective by which to assess current and potential museum audiences and the programs that museums can develop to appeal to and satisfy various sizable groups. If we are to reach those who are not coming frequently or at all, it is essential to program for more than one type of audience. Each attendance group is looking for different types of benefits in leisure experiences. Frequent visitors—the smallest group—are, for the most part, finding what they want in museums. But for the occasionals and the nonparticipants, who seek an opportunity to interact with people and to relax, the prospect of going to a museum for a learning experience, for a challenge, for doing something worthwhile in leisure time, is not enticing. Particularly if these people have had negative experiences with formal education, the idea of going to a museum for a learning activity connotes an exacting, ponderous undertaking rather than an enjoyable, casual experience.

If we museum professionals are concerned about reaching new audiences—the occasionals and the nonparticipants—we must appeal to them on the basis of what satisfies *their* criteria of a desirable leisure experience. Endeavoring to reach these elusive audiences by providing more of the same type of programs now offered, regardless of their quality, will not pay dividends for either audiences or museums; a different emphasis and presentation are necessary.

For instance, instead of portraying itself as an educational institution, where the family learns together, the museum seeking to reach occasional participants and nonparticipants might stress that it is a place for exploring and discovering, for enjoying a relaxed family outing, and for having a good time with other people.

Discovery workshops that offer the family the opportunity to participate as a unit—to identify insects, fungi, or fossils in a natural history museum, to work with clay or on a mural in an art museum, to try on or make facsimile costumes in a history museum—are examples of current museum programs that occasional visitors prize. These activities, though, should not be just ends in themselves but utilized as entrees, transitions, into the collections. If skillfully handled they can prepare the occasional visitors to cope with the "museum code" as well as enhance their positive perceptions of museums as places that meet their criteria of satisfying leisure locales. If they find their preferred attributes are present, they will *choose* to return. Other museums are providing tours and talks geared to the interests of specific groups—construction workers, sports fans, hobbyists of all hues—to demonstrate the relevance of museum collections to persons who do not perceive any connection between museums and their lives.

None of these approaches implies that the museum will abandon its current purposes or programs. They do require that museum staff and trustees view exhibits and programs from a different perspective, presenting them in as many appealing manners as possible.

Before we solicit the nonparticipants' attendance, therefore, we will have to consider what the uninitiated expect in the way of assistance with the "museum code." Are adequate helps provided so that those who make the initial venture onto untried turf will receive enough benefits to want to return? This does not mean diluting the message, but it does mean communicating the message in the nonparticipants' terms, on several levels of detail and comprehension, in order for them to perceive it as meaningful to their lives.

Gathering quality information for long-term decision-making is worth effort, time, and money.

In addition, it is essential to remember that occasional participants weigh each museum visit against other leisure options. They may choose to attend on a particular occasion *instead* of watching television, browsing at a shopping mall, participating in a sport, or working in the garden. A museum outing for them is likely to be a vehicle for having an enjoyable time with other people rather than for concentrating on the content of the exhibits. While participating, they hope to find comfortable surroundings in which they can feel at ease, both physically and psychologically.

Applying these findings and doing follow-up studies can benefit all museum professionals by helping to build a fund of reliable information about audiences. Although a major study of similar scope cannot be accomplished without expertise in audience development and scientific research techniques,

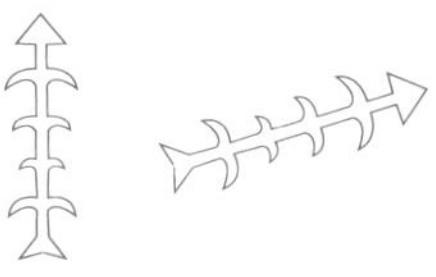

each museum can utilize systematic procedures to obtain psychographic information on its audiences, on their perceptions of the museum, and on ways to deal with practical situations such as communicating more effectively with a variety of publics. Gathering quality information for long-term decision-making is worth effort, time, and money, for the value received is in direct proportion to the care invested in designing and carrying out a carefully controlled study.

The Toledo Museum of Art has benefited itself and its audiences by incorporating the findings of this study into its planning and programming. Most important has been a heightened awareness by the staff of the diversity of visitors' expectations, leisure values, and needs.

The primary influence of the research results on the museum's major structural renovation was in recognizing the need for public amenities. People need a sense of where they are in relation to the whole museum and to other areas in it. They need to feel comfortable. New graphics greatly improved visitor orientation, and comfortable seating was added to the new entrance lobby. An information office to welcome visitors was located just inside the entrance.

For the recent *El Greco* exhibit, the museum targeted the area's Hispanic and Greek communities for special attention. Both groups attended in greater numbers than might have been otherwise expected, especially the Greek residents, who constitute a small, closely knit community. Still under way are efforts to improve labeling and explanatory handouts to assist the less-prepared visitors in understanding why certain art works are grouped together.

Most of all, the study results prompted the staff to think beyond only internal concerns and "to consider the public we're doing this for," explains Gregory Allgire Smith, assistant director for administration. "We now realize we should plan more on their terms, not just on our own terms."

We, too, can solve many of our audience development problems if we recognize that occasional participants and nonparticipants are looking for experiences and rewards different from those they now find in museums. If we want them to love museums, we must offer them some of the values that are important to them, in programs that meet some of their needs, while we continue to provide what the frequent visitors already find satisfying and rewarding.

N o t e s

1. Marilyn G. Hood, "Adult Attitudes toward Leisure Choices in Relation to Museum Participation" (Ph.D. dissertation, Ohio State University, 1981)

Marilyn Hood, former editor of special publications at the Ohio Historical Society, is a museum consultant specializing in audience development, evaluation, and market analysis. She lives in Columbus, Ohio.

Works on audience research suggested by Marilyn Hood are listed in a separate bibliography on pages 152-153.

CASE STUDIES

Marcia Howe describes how the Carter House Natural Science Museum conducted research on the perceptions and preferences of its audience, and used the resulting information to plan for growth. Drawing on her experience at the Brooklyn Children's Museum, Suzanne LeBlanc analyzes what it takes for an institution to succeed in diversifying its audience and gives examples of successful tactics.

AUDIENCE RESEARCH TO GROW ON *Marcia R. Howe*

Carter House Natural Science Museum was opened in 1977 in Redding, California, as a small junior science museum. The museum board and director decided in 1985 to begin officially the planning for a new facility. Since its opening in a modest renovated home in a city park, the museum had been bursting at the seams. However, with its minimal beginnings, it took nearly nine years to assemble the strength and backing to undertake this major task.

The development of a new science museum project evolved into a joint project with a consortium that planned to build a museum park. The size and scope of the joint project pushed into the political arena. In addition, the site for the new facilities began to be questioned by some in city government.

In the midst of this highly politicized milieu, the museum undertook a feasibility study in two parts, one of which evolved primarily into an audience research study. The board and staff realized such a study was important whether or not we were going to build a major new facility.

We had several concerns that we felt should be examined. Two of these related to identity: Were we being perceived primarily as a children's museum? And were we being confused with a neighboring museum? Additionally, we were in a position of changing and growing, and felt that information on our image and position in the community was essential for effective planning.

THE STUDY PROCESS
We contracted with a firm with whom we had previously worked on board training. The firm was located in San Francisco, 200 miles to the south. The fee for the study was $10,000, with an additional $2,000 in expenses.

The goal of the audience research study was to provide information that would help the board and staff learn about the identity of the museum within the Redding community, and examine the implications for future development and decision-making. The specific objectives of the study were:

1) to determine a demographic profile of the groups who use the museum and its services

2) to identify potential users who could be attracted to support the museum and its programs

3) to examine programs and activities within the current space and facilities

4) to determine the image that the museum currently portrays in the community and whether barriers to visitation exist

5) to address members' perceptions about current programs and value received from supporting the museum

6) to analyze how people prefer to use their leisure time and what implications these trends have for the museum

7) to survey public opinion regarding the future of the museum

METHODOLOGY

The museum staff and board worked with the consultants on the survey instruments. Questions were submitted to the consultants. They combined them with their own and then gave the surveys to us to review. Also in the study were several questions calculated to obtain community residents' opinions about generating economic support for cultural institutions in Redding, which might address one issue of relocating the museum. The questionnaires were then pretested by staff and community participants.

The main staff involved in the project were the museum director, who helped develop the questionnaire; clerical staff, who assisted in the mailings and processing of the returned forms; and museum volunteers, who participated in the research.

Data was collected in several ways. Our consultants held interviews with community leaders who gave background information on Redding, and an idea of the community's awareness of the museum's accomplishments and future plans. Our group of community leaders included the president of the Chamber of Commerce, the city manager, the president of the Motel Owners Association, the city planning director, and the director of the county Economic Development Corporation, as well as directors of arts and tourism organizations.

A membership survey was sent to the entire membership of 350. The questionnaire asked for responses to membership benefits, programs and exhibits, preferred leisure-time activities, members' perceptions of Carter House, and demographic information. In addition, a section was provided for members to express their opinions about issues affecting the future of Carter House, including the proposed relocation. The questionnaire was sent with a gift of postcards and a cover letter from the museum director to encourage response. Approximately 40 percent of the membership responded to the survey.

Comments from "interested citizens" were solicited from a random sampling of the city's utilities customers. Local residents received questionnaires about their perceptions of Carter House and its programs, their opinions about the future, and demographic information. As with the membership, interested citizens were sent a small gift, a museum decal, and a letter from the director. The Redding Electric Utility department assisted in selecting a random sample of participants for this part of the study. Of the 21,907 homes that were billed for utilities, 2,434 residents received questionnaires, or every eighth resident listed on the utilities roster. The return rate was 9 percent, which is very high for direct mail, where 3 to 5 percent is considered a good return. However, from a statistical point of view, the results had to be analyzed keeping in mind that interested citizens who used the museum were more likely to respond than non-users.

Therefore, in order to validate the community questionnaire responses, 150 additional community interviews were conducted by volunteers and state university students. A total of 11 museum volunteers and one staff member (our store manager) worked on the interviews. A professor of museum studies at Chico State University, 60 miles to the south, volunteered along with three of his students. In addition, eight volunteers from a neighboring museum assisted. All were trained in two sessions with the consultant.

A controlled setting for these interviews was not possible. Initially we planned to conduct the interviews at one shopping mall, but we discovered a special permit was required. Due to the time factor of the study and the availability of the volunteers, the interviews were held at supermarkets or wherever the volunteers felt a variety of adults might congregate. Since the method of interviewing was less controlled than originally planned, the information was not used throughout our results, but was incorporated as back-up to key findings.

The research volunteers also interviewed 99 museum visitors over a period of four weeks at various times during the day and week. The visitors were asked to comment on the highlights of their visit and areas for improvement.

WHAT DID WE LEARN?

- We learned that most of our members were "joiners." They joined primarily to "support" the museum, not for any additional benefits. While many other people were interested in and visited the Carter House, they saw no reason to join.

- To no great surprise, all the data indicated that the live animals are the part of the museum people like best.

- We learned that people were well-served by our hours,

and that expanded evening hours were not necessary.

- We confirmed that we needed to do some work in public relations. Many people had not heard of the museum. Also, we were perceived primarily as a "kids only" place. At the same time, we learned that many "interested citizens" had no children living at home, and thus we might consider this group as a potential audience.

- Another indication of a problem with image and identity was the fact that nonmembers often confused the museum with the nearby art and history museum.

- We confirmed that most respondents place a high priority on outdoor recreation.

- The majority of respondents indicated agreement that continued support by Redding of its science museum was important. The respondents were very enthusiastic about the type of cultural development envisioned for the museum park project.

- Questions about bond assessments and tax increases to support the museums received mixed responses, but a small entrance fee was supported by most respondents.

HOW DID WE USE THIS STUDY?

First of all, the study results supported several initiatives we had already taken. We had developed an adult-oriented program to appeal to non-child audiences, and we had developed a new logo that was graphically sophisticated and definitely appealed to adults as well as children. The importance of these changes in developing a new image was confirmed.

We increased our focus on membership. The study alerted us to the fact that no real benefit of membership or distinctions between members and nonmembers existed. We began to put more energy into maximizing benefits such as the newsletter, and member discounts and freebies.

A few years later we instituted an admission fee. The favorable response to this idea indicated by the study helped us overcome objections to this important move to increase operating income and to provide free admission as a major membership benefit.

WHAT WERE THE BENEFITS OF THE STUDY?

The study confirmed some of our observations and gave us a fresh perspective by seeing the museum through the eyes of a variety of people from the community.

The particular type of interaction with community leaders was very helpful politically. In our community of 60,000 residents, public officials and others are extremely impressed when an organization undertakes a project as sophisticated as marketing or audience research. In addition, our consultant was helpful in conveying an outsider's view of the museum to these leaders, as well as information about our future plans.

The study helped us identify some key areas in which we needed to do more with public relations and membership. We have had much success in both areas as a result.

The study gave us a baseline profile of our audience. When planning new exhibits, programs, or a membership drive, we know with whom to start for input. This baseline gives us an idea of the progress we are making in expanding the audience.

In addition, as we develop our new facility we will need to develop a marketing plan. The study gives us an idea of not only the existing audience, but also potential audiences. We can ask such questions as where do we want to go in terms of increasing the commitment of existing visitors and developing new audiences. For example, since outdoor recreation was confirmed to be such a popular area of activity for Redding adults, programs in that area such as nature walks, nature photography, and fishing-related activities would be productive areas of expansion.

EVALUATION

Although the study was not statistically stringent, we feel it provided a fairly accurate picture of existing and potential audiences. It definitely gave us a basis for developing programs, exhibits, and the overall management of the museum. The study would have been more helpful if we had been able to have some focus groups with educators and if we had been able to get information from summer tourists, but the timing of the study prevented this. Nonetheless, the board and staff gained a valuable perspective about what the museum means to the community and what types of expanded programs and facilities would be supported.

Marcia Howe has been on the staff of Carter House Natural Science Museum since 1977. Originally a part-time museum coordinator, she is now executive director. Ms. Howe has a B.A. in history and social sciences, and is a graduate of the Museum Management Institute. She has been active in the California Association of Museums and the Northern California Association of Museums, and is a founding member of the National Association of Interpretation.

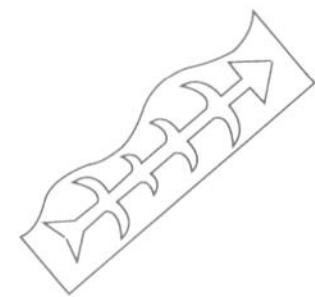

REACHING THE UNDERSERVED
Suzanne LeBlanc

So you want your science center to develop an economically and culturally diverse audience. Granting agencies emphasize its importance. Professional conferences highlight the issue in numerous panel discussions. You've had some ideas and tried a few of them out, but nothing has really worked. You've held a free night and no one came—except all the people who can afford to pay. You're not sure what to do next, or why what you've tried hasn't been successful. But you are still committed.

The following thoughts, suggestions, and examples may help. They spring from work done at three institutions: The Children's Museum in Boston, The Brooklyn Children's Museum, and Lied Discovery Children's Museum in Las Vegas. The audience issues at each of the museums vary significantly. But all share an institutional commitment to audience diversity, an openness to collaboration and community involvement, and a willingness to commit staff and resources to the effort.

The process of developing an economically and culturally diverse audience requires a thoughtful and comprehensive approach, and an understanding that progress will be slow and incremental. Before beginning, it helps to consider some of the reasons why some people might not visit a museum:

- They may not be able to afford the admission price and/or transportation cost.

- They may not feel welcome when they do visit, not seeing many people who look like them visiting or working there.

- They may be afraid that they won't know what to do when they get there.

- The program and/or exhibition content may not be relevant to their lives or may assume a high level of prior knowledge about the topic.

- They may never have learned how to use a museum.

There are many examples of programs, techniques, collaborations, and outreach efforts that have helped cultural institutions be more accessible to a broader audience. Whether or not these work in a particular institution often depends on frequently overlooked and more subtle considerations such as institutional commitment, staff training, and staff diversity.

INSTITUTIONAL COMMITMENT

Success at attracting a diverse audience to your institution will require a concerted effort over a long period of time. This will only be possible with a strong, articulated institutional commitment to the goal. An articulated policy, reflected in your mission statement and long-range plan, will help ensure that when resources are tight, all efforts do not come to a halt. Support from the board of trustees, the museum director, the development and public relations offices, and staff throughout the institution will ensure that audience diversity becomes an integral part of the museum's agenda.

Attempts at diversity should include diversifying the board of trustees. As the governing body of the museum, trustees represent the museum in the community at large. A diverse board will send messages to the community about institutional concerns and priorities.

STAFF TRAINING

All staff who come in contact with the public, and ideally all staff, should receive training in how to make visitors feel welcome. This will enhance the museum experience for everyone, but is especially critical for first-time visitors and those who feel unsure about the museum. Staff sensitivity training is a basic but poorly understood element of any audience-building program.

It is very important for management to clearly communicate, through policy, training, and example, that the museum is committed to welcoming a wide range of people to the institution. Staff who deal with the public day after day will pick up and communicate, for better or worse, the philosophy, attitudes, prejudices, and beliefs of the institutional culture. Clear, strong leadership and vision from management will be required to set a tone in the museum that accepts, values, and welcomes a diverse public.

Specific workshops on topics relevant to audience diversity will help inform and sensitize staff and are frequently offered at museum and educational conferences.

STAFF DIVERSITY

An active, committed effort to diversify staff is integral to any attempt to attract diverse audiences. Think for a minute about any experience you may have had visiting some place where you felt like "the other." How did it make you feel? Welcome or unwelcome? Comfortable or uncomfortable? Having a diverse staff sends messages that say to a child or to an adult, "I am welcome here. There are people like me here. I belong here."

Success in diversifying staff requires consistency in the application of policies and procedures across the institution and strong advocacy. If hiring is decentralized at your museum, it is particularly important to have a clear statement in the personnel policy handbook that expresses the philosophical commitment of the institution and delineates practical procedures to guide the process.

SOME PROGRAMS THAT WORK

Free Nights

Holding a free or low-cost admission night (morning, afternoon) seems like a sensible way to begin attracting visitors who would not come because of a prohibitive admission charge. The Children's Museum in Boston has been open on Friday nights either free of charge or with a one dollar admission fee for about 20 years. Their experience is instructive for other museums considering such a program.

When the program was first begun, it was advertised in the daily newspaper's calendar of events. At first hardly anyone came; it takes time for the public to integrate a change in hours. Then, when the audience for Free Friday Night began to build, it was primarily composed of people who could afford the full admission price. But the museum did not give up on the idea of using Free Friday Nights as a tool to broaden its audience.

The museum decided to host schools on Free Friday Nights where museum staff had participated in curriculum and teacher-training projects. So, at the end of a year of working with the inner-city McCormack School, the museum hosted a McCormack School Night. Grants paid for bus transportation and refreshments. The museum did the same thing with its community work. After doing outreach in after-school programs in Chinatown, for example, the museum hosted a Chinatown Night, again with free transportation and refreshments.

This accomplished a few very important things. People who had never been to the museum found a "safe" way to visit for the first time. There was a "reason" to go—the special school or community night. The trip was made in a group, in a familiar context (people knew other people, or knew the school or community center). Transportation was provided, which made it easier the first time around, even for people who had a car or lived near public transportation.

For people who didn't know that the museum was open free every Friday night, the community or school night provided a comfortable way to find out about and experience it. The museum was open to the general public at the same time, so it was clear that those who came with the special group could return for free any Friday night on their own. The museum made every effort to provide an especially welcoming and friendly experience with signs, refreshments, and staff present who had worked with many of the kids at their schools or community centers. Over time, Free Friday Night was very instrumental in changing the composition of the museum's audience.

- Do not expect a free night to attract a new audience overnight.

- Think about using a "free night" to celebrate your museum's work with a particular school or community group.

Community Involvement

There are a variety of ways for a museum to involve new audiences in a meaningful way. Effective collaborations with community agencies, leaders, and schools can open many doors that would otherwise be closed. Collaborations allow a meaningful dialogue to occur that builds solid relationships, enhances feelings of goodwill, and allows the expression of diverse perspectives and points of view.

Advisory Boards

Advisory boards provide an opportunity to get input about specific issues, programs, or projects from experts and advocates. Museums have assembled advisory boards related to audience diversity in such areas as making museums accessible to people with disabilities; how to attract, program for, and serve teenagers in museums; and how to make a museum's programs and exhibits more reflective of diverse cultures. The relationships formed between the museum and its staff and advisory board members have the potential to last long beyond the life of a particular project.

Collaborative Programs

Collaborative programs, especially ones that have the ability to become multi-year projects, can not only introduce new audiences to the museum, but can also turn them into confirmed museum-goers.

The Brooklyn Children's Museum, interested in doing more family programming, participated in a model collaborative program with the New York City Department of Cultural Affairs, the Human Resources Administration, and the Board of Education. The audience for the program was families with children under five years old living in transitional housing. Families came to the museum once a week to see one exhibition each time. They would then spend an hour and a half with two museum educators, doing activities focused on interaction between parent and child—songs, crafts, storytelling—and related to the exhibition content. The activities were things the families could readily continue on their own. Family groups also made something to take home, tying exhibit content to the circumstances of their lives.

A number of factors contributed to making this program successful. New York City officials trained museum staff so that the program design did not demand the impossible (e.g. several days per week attendance rather than one). Funding paid for adequate time for program development and preparation and allowed two staff to work with the parents and children. Encouragement and support for participation was provided by city personnel at the shelters.

As families left the shelter for permanent housing they asked if they could continue participating in the program for their full 10 weeks. And at the end of the program parents asked if they could continue visiting the museum on their own.

While the number of people affected by the program at The Brooklyn Children's Museum was small, many museums throughout New York City participated. The number of children living in transitional housing who were involved with museums through this program was large. And the lessons learned about the ability of museums to provide a meaningful experience to children in difficult circumstances were not soon forgotten.

PROGRAM IMPLEMENTATION

There are many sources of funding for museums wishing to serve audiences not traditionally served by museums. Corpo-rations will often sponsor a series of free nights at a museum. Private and government foundations will fund programs geared to audiences and topic areas that fit their focus, such as adolescents, low-income audiences, and girls in science. Some foundations that do not usually fund museums will support these kinds of programs at museums and science centers.

A number of practical matters can influence an institution's ability to carry out programs to reach underserved audiences.

- It is beneficial to have someone on staff who advocates for audience diversity. This can be a funded outreach coordinator or just someone on staff who cares about the issue.

- It is important to broaden your public relations effort to include community newspapers and radio, flyers posted in community agencies, etc. Think about doing some bilingual or multilingual flyers.

- Don't forget that economic diversity and cultural diversity are not identical issues. Reach out to people whose upbringing did not include learning to use museums, not only to people from various ethnic backgrounds.

Suzanne LeBlanc has worked for children's museums for nearly 20 years, including The Brooklyn Children's Museum and The Children's Museum of Boston. She has served in a variety of positions, and issues of audience diversity have always been a primary concern. She now serves as executive director of the Lied Discovery Children's Museum in Las Vegas.

PLANNING
EXHIBITS &
PROGRAMS

With your mission established and audience identified, you proceed to define the exhibits and programs that will offer visitors the kinds of experiences you intend. You will face issues of content, style, and staffing that need resolution before displays can be built and programs launched. This chapter deals with these organizational issues. It begins with a discussion of themes—the messages your exhibit collection, supported by programming, is meant to convey—and then discusses strategies for developing exhibits and for establishing programs.

SELECTING CONTENT

All of science won't fit into any one building, so you need to develop some sort of rationale for selecting the phenomena and scientific ideas that exhibits and programs will illuminate. You will probably consider many factors, including visitor appeal, available scientific expertise, funding, and educational payoff. To determine what your institution should be about, you may wish to brainstorm with other stakeholders—educators, local industries, constituent representatives. Ultimately, a single vision of the institution's content and methods needs to emerge that an executive director, or whoever is acting in that capacity, can implement.

Some science centers include a content focus in their mission, such as health sciences. Most centers are more general, claiming physical science, life science, and applied technology as their bailiwick. Within such broad boundaries, content choices have to be made. Some science centers select themes under which they group a number of related subjects, with exhibits and programs devoted to each subject. Other centers take the opposite tack, providing an arcade atmosphere of miscellaneous interaction. Some centers select subjects with a storyline, such as human reproduction. In this case, their communication goals may include conveying a body of information, as history and natural history exhibitions often do, and so may imply use of traditional museum methods as well as experience-oriented ones. Some centers mix scientific and technological subjects within a comprehensive theme.[1]

Do visitors prefer one theme over another? Do visitors notice themes and use them to organize their visit, or to make conceptual sense of their experience? Most of the research in the field focuses on the effectiveness of individual exhibits; there is very little systematic information on how science center visitors respond to a collection of exhibits on a given subject or to still more general themes. Watching visitors explore unfamiliar materials at the most concrete level and in no particular order, many old hands in the science center field believe that themes are not generally salient, or that we haven't yet found ways to make them salient for visitors. But themes help planners define their institution, organize its content, and communicate about it as they search for support.

To select content, many people planning science centers consider what is unique in their setting, what competitive attractions offer, what prominent supporters care about, what they imagine or, better, ascertain their public cares about, and what they themselves find fascinating. Eventually, it boils down to personal choice, informed by knowledge of the exhibit medium. Many subjects lend themselves to hands-on investigation; that is, they offer intrinsically interesting phenomena and ideas that can be successfully, repeatedly probed by members of the public given appropriate apparatus and assistance. A new center will do the best job with subjects in which key people—the scientists and educators who function as staff—feel the keenest interest and competence.

One way to be sure that a new science center succeeds is to choose exhibits and programs that "work" elsewhere. Many

new science centers have done this: They have adapted individual devices or sets of exhibits or popular programs from leading centers, particularly on the subjects of mechanics, optics, perception, bubbles, computers, and the human body. So-called "cloning" saves a great deal of time and money because the R&D has already been done. Many leading science centers encourage cloning by publishing exhibit and program descriptions, giving workshops on technique, and even building copies of exhibits for others.[2] A handful of places will sell ready-made exhibits to other centers. But successful interpretation of cloned material still requires skilled and knowledgeable people at the new center.

Inventing exhibits and programs creates an ambience of experiment and discovery that permeates and enlivens an institution.

Cloning may not be advisable in locations where visitors can choose among several science centers within driving distance. A stronger disincentive is the role cloning plays for the staff. Energetic, committed people—the kind a new center wants to hire—may not be satisfied with interpreting other people's exhibits and programs. They will want to realize ideas of their own. Inventing exhibits and programs, moreover, creates an ambiance of experiment and discovery that permeates and enlivens an institution. Recognizing the value of such an ambiance for both staff and visitor, many new centers decide to offer a combination of cloned and original exhibits and programs that correspond to their intended message.

Visitors respond to interactive exhibits on a wide variety of subjects; the subject itself would seem to be less important than its treatment.[3] While the literature about science centers doesn't argue for selecting one subject over another, it does suggest that the number of subjects addressed be small. Taking their lead from cognitive psychology and constructivist education,[4] science center researchers have begun to investigate the way in which visitors' preconceptions, so-called "naive theories" about the way things work, affect what they make of exhibits.[5] Researchers closely observe and question visitors about their thoughts as they handle displays. As expected, visitors' notions are powerful and obdurate, constraining their response to the material at hand. In order for visitors to extend their thinking to incorporate new elements, the research implies, they need to confront a particular phenomenon or notion many times. Redundancy makes for learning. It also requires floor space and programming time. To do justice to a subject, or rather, to do justice to the visitor approaching the subject, a science center should devote considerable space and resources for presenting it. Practically speaking, that means limiting the number of subjects treated.

In a parallel line of reasoning, many science center staff argue for interpreting a given subject through the lenses of many disciplines. The Exploratorium has strongly advocated using art as well as science to illustrate phenomena. Some of their most popular and widely emulated exhibits, like the touchable tornado, are the works of artists. Other centers integrate mathematics in displays, and still others use historical perspectives (like artifacts and dramatic recreations) for purposes of comparison. Surely, the richer the mix, the more people will find something attractive and resonant in it. But more expertise, time, and money will be required to mount a rich, multidisciplinary exhibit collection.

James Bradburne and others argue that science centers should not "present science as a collection of demonstrated principles" but rather show it "as an on-going process in which one could genuinely participate."[6] Bradburne argues for "a second-order experience," so visitors leave not saying "I know" about a particular phenomenon, but rather "I know how I know." Many science educators, reacting to schooling that emphasizes facts and vocabulary, agree that the active doing of science by real people needs emphasis. They point out, also, that as scientific knowledge grows geometrically, only people who understand the scientific process can pick their way through it.

Science centers should open a window on an aspect of nature and help visitors investigate it as scientists do.

But educators also warn that process skills cannot be taught in a vacuum. Observation is often defined as a key element in the scientific process; but no one observes just to observe— you observe in order to see something to which you want to pay attention and about which you might want to do something. Furthermore, the nature of the observation flows directly from the subject matter under investigation. Observation can't be separated from the intellectual context of scientific work. Educators urge us to "help students acquire scientific knowledge of the world and scientific habits of mind at the same time."[7] Science centers should open a window on an aspect of nature and help visitors investigate it as scientists do.

In practice, exhibit planners should intertwine content goals, process goals, and the visitor's need for comfort in designing displays. Ideally, each display states clearly what it is about, in words and by its physical design; reveals an intriguing phenomenon; invites the visitor to handle it in a structured way

that makes for a successful interaction; and allows for free experimentation along several dimensions. The collection of displays—and related programs—on a given subject provides many different access points to the material and redundant yet engaging experience. Of course, most visitors won't explore a subject systematically, so every exhibit must stand alone. But over the course of many visits, or in a staff-guided tour, visitors may take in a good deal about a given topic.

DEVELOPING AN ENVIRONMENT

People planning science centers sometimes ask about the effect of architecture on visitors' ability to learn from displays. Does an open plan or a series of enclosures encourage exploration and learning? Some science centers prefer a wide open space, which begs for exploration yet lets visitors know where they are in relation to the whole. Other science centers prefer rooms or smaller galleries in which visitors can readily achieve a sense of closure. Going from room to room, then, provides a visually refreshing break.

The literature on environmental psychology suggests that both positions are valid. Vast or intimate parts of a science center can perform different functions, depending on the content and nature of the experience envisioned. If visitors need to concentrate, they may be more comfortable in a cozier space, off the beaten track. On the other hand, some apparatus works best on a grand scale. Two studies on the use of audiovisual displays suggest that seating can be an important consideration.[8] For science centers that can construct their own space, form can follow content.

One last note on architecture: Most science centers regularly mount temporary or traveling exhibitions to attract repeat visitors, offer new content, and remain in the public eye. People planning new centers generally set aside a temporary exhibition area or gallery commensurate in size with the rest of the institution. Sometimes they also set aside space to show works-in-progress to visitors in order to preview their response.

Mounting your own temporary exhibitions is a wonderful outlet for staff creativity; it's also an operational burden. To ease the burden, many centers borrow exhibitions from others. In the U.S., both ASTC and the Smithsonian Institution operate traveling exhibition services that rent science exhibitions for a range of fees. A number of science centers also circulate traveling exhibitions, and other centers belong to consortia to share exhibitions developed by members. Attending the ASTC annual conference is a good way to find out about such opportunities.

DESIGNING EXHIBITS

In 1986, Tony Sin, then at the Ontario Science Centre, studied the center's temporary exhibition on food to find out what makes individual exhibits successful. He showed photographs of the 51 displays to a sample of visitors and asked which 10 should be included in the center's permanent collection. He then ranked exhibits in order of visitor preference and compared their characteristics in terms of exhibit design.

Sin discovered that some things don't matter to visitors: the presence of computers or push-buttons; the cost of materials or the exhibit's level of finish. Some things do matter: easy access on the exhibit floor; frontage between seven and 22 feet (focused, but with room for a group); and inclusion of live plants, animals, and demonstrations. Some things matter even more: whether an exhibit's subject matter has intrinsic meaning for visitors themselves; whether the display provides feedback. The single highest correlate of popularity, though, is the amount of staff time devoted to researching and designing the display.[9]

Sin's work can help answer a question often asked by people starting science centers about the level of polish exhibits should have. Exhibits in existing centers show a wide range, from "put-together with string and rubber bands" to "corporate high style." Some people believe that high polish and special effects get in the way of learning because visitors won't identify with the exhibits, although they may admire them. Others say visitors in certain institutional settings expect and welcome polish. Sin's work confirms that there are no hard and fast style guidelines. Visitors' liking of exhibits depends not on the exhibits' level of finish, but rather on their meaning and the feedback they provide.

Building an interactive exhibit requires work on at least four dimensions: science, communication, engineering, and aesthetics.

Why should the most popular exhibits take longer to create? Consider that building an interactive exhibit requires work on at least four dimensions: science, communication, engineering, and aesthetics. Someone who knows and loves science must point out intriguing phenomena and ideas worth displaying. Someone who understands the audience's needs for cogency and clarity must insist on describing the subject in visitors' terms. Someone with engineering (and sometimes computer) skills must ensure that the device works safely, pleasantly, and reliably. And someone skilled in visual arts or design must configure the exhibit attractively and mount it effectively on the exhibit floor. It's a rare soul who can handle the whole job, so it is usually done by a team of people. The more time team members spend talking together, sharing perspectives and enthusiasms about the subject, and clarifying their objectives, the more likely they are to get an exhibit right—to capture the imagination of visitors.

Many science centers, starting with limited budgets, can't afford to employ a full-time team. Sometimes they rely on skilled volunteers to clone or build new exhibits. Sometimes they employ exhibit design firms to develop a given piece of work at a defined cost. However an exhibit is built, expertise on all four dimensions needs to be assembled or the work will fall short. And someone needs to be in charge, to keep up the momentum and resolve the inevitable differences among members of the team.[10]

There may be many different people involved in making exhibits at different times as a center grows. A new science center does well to articulate its communication goals and exhibit development methods on paper for all the various players, and to perform as much exhibit development work as possible in-house in order to build institutional memory and skill. At first the executive director (or person acting in that capacity) should oversee the development process, holding the team to the mission and vision as defined. He or she will need to allow for surprises and false starts, since exhibit making remains an uncertain art.

Once a subject is selected and an R&D team assembled, the team's first task is to decide what is interesting and exhibitable about the subject and compile an initial exhibits list (from their own experience, books, teachers, other centers, friends). In this crucial conceptual stage, the team sometimes queries sample visitors and other staff to hear their questions and conjectures about the topic. Touching base with the audience can lead a team to reject a topic or revise its approach. It can help the team decide what to include, what to ignore, which misconceptions to combat, and how to conceive specific displays.

A new science center does well to perform as much exhibit development as possible in-house in order to build institutional memory and skill.

In this pencil-and-paper stage, it is useful to ask what you expect the visitor to do at a given display and what you think he or she will get out of it. The display has to offer a "payoff" to reward the visitor's attention and effort to deal with something new. Payoff can be a surprising observation, a delightful sensation, a meaningful association, the satisfaction of having mastered a system, or, best of all, a combination of these. Trying out an early idea on a few people, quickly and cheaply, can give you an idea whether it will, in fact, pay off, preventing you from making mistakes that cost more time and money to rectify further down the line.

Once exhibits have been specified, builders develop the designs and make mock-ups of new or complicated displays for additional testing. These prototypes, perhaps rough-hewn and hand-lettered, are complete enough for outsiders to use. In an operating science center, prototypes can be placed on the exhibit floor—visitors generally enjoy helping to develop new displays. The design team observes and questions visitors, looking for durability, ease of use, appeal, and comprehensibility of the device and its explanatory text. The team also checks how people of different ages and backgrounds use the display, noting areas of confusion. Feedback from the testing is used to revise and improve the displays before they are fabricated in final form.

Testing prototypes, also called "formative evaluation,"[11] can add 10 percent to the cost of building exhibits and require extra time, but it provides the only assurance that an exhibit will work, for most of the people most of the time, on the exhibit floor. Even the most experienced and artful exhibit builders cannot anticipate how various members of the public will respond in detail to the actual exhibit and its controls, and they routinely exercise the discipline of checking with the audience and modifying accordingly.[12]

Most design firms are not accustomed to planning and budgeting for prototyping, which argues for accomplishing at least some of this work in-house. When an idea has been tested and adjusted, it can then be efficiently drawn up for manufacture by in-house or outside fabricators. Once fabricated and installed, exhibits may still require some adjustment (the "burn in"). At this stage, revision is extremely expensive, and most people avoid it.

Sometimes staff collect information on how visitors respond to completed exhibits (called "summative evaluation") to assess accomplishments and derive lessons for future exhibit development efforts. Quite a few funders want science centers to monitor their progress and seek more objective measures of visitor learning. Although there is a growing amount of literature on summative evaluation of exhibits at science centers, there is as yet no consensus on optimal techniques.[13]

Exhibit Facts and Figures. Several designers of interactive exhibits have written useful guidelines for exhibit development and construction.[14] These books contain practical advice for solving many design problems. For general planning purposes, you can use the following rules of thumb:

- A typical exhibit requires 100 square feet.

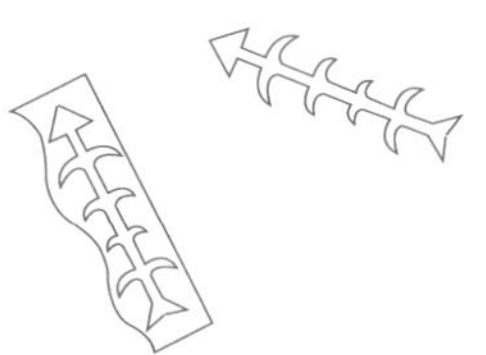

- The average cost of finished exhibit space is about $200 per square foot (including new exhibits and walkways around them).

- Developing an individual display from scratch can cost anywhere from a few hundred dollars (mirrors) to a hundred thousand dollars or more (interactive videodisc).

- The effective lifetime of an exhibit is five to seven years.

- A crew of 10 people devoted exclusively to exhibit building, with access to machine shop, electronics, and graphics equipment, can plan, prototype, complete, and install about a dozen original displays in a year. Or, the work can be parceled out among many contractors to speed completion.

- In the case of very small, low-traffic science centers, exhibits can be created much more inexpensively, using volunteer labor and donated materials.

Exhibit maintenance is a fact of life at science centers, all too often underemphasized and underbudgeted. Even the best displays wear out with steady usage, especially when visitors do things the design team never anticipated. Nothing useful happens to visitors at non-functioning displays, and sometimes their frustration even leads to further damage. Consequently, people try to remove broken exhibits from the floor as soon as possible. Unless someone is repairing displays regularly, the visitor winds up with less and less to see and do.

To help repairs happen, exhibit designers should keep maintenance considerations in mind; displays should disassemble easily, be easily cleanable, and rely on interchangeable parts whenever possible. It's hard to say how many person-months of exhibit maintenance work a center will require because much depends on the density and complexity of its exhibits. Maintenance and design interrelate: When you fix a broken exhibit you have a chance to improve it. In centers where staff are actively involved in on-going exhibit development, there's an institutional stake in "getting it right" that promotes attention to maintenance and sometimes results in developers and maintainers sharing the load.

> **In centers where staff are actively involved in on-going exhibit development, there's an institutional stake in "getting it right."**

Safety and accessibility issues also come into play in designing exhibits. Safety requirements have not been codified for science centers, so each center must establish its own procedures.[15] In general, exhibit development should include a safety review informed by past experience. A center needs to keep careful records of injuries to visitors and to articulate a clear policy on how injured visitors and their families should be treated, both medically and compassionately. All staff members need to learn the part they are expected to play in keeping visitors safe. Accessibility issues have been studied for science centers and a variety of guidelines have been published.[16]

LAUNCHING PROGRAMS

Exhibits cost a lot to build, but once installed, they can be maintained with a modest outlay. Programs presented by skilled staff, on the other hand, may not require a large capital investment, but they require salary commitments and often serve fewer visitors than exhibit halls. While an exhibit project can be canceled if fundraising falls short, it may not be possible, let alone desirable, to dismiss employees. Start-up institutions try to avoid burdening themselves with salaries, yet most of them invest to some degree in people who develop and present interpretive programs to visitors. Like other science centers, they believe that programs can accomplish mission-related goals that exhibits alone cannot.

Staff at leading science centers often say that some of the best experiences their centers offer visitors come from interaction with the people who work there. Even casual conversations with people changing light bulbs on the exhibit floor can lead to memorable interaction. Staff who are versed in science and skilled in communication can answer visitors' questions and tailor comments to visitors' personal interests and backgrounds, helping them explore subjects more deeply than is otherwise possible on the exhibit floor. Staff can model exploratory behavior and science process skills in engaging, non-threatening ways. They can also serve as role models for underrepresented audiences, if they are members of those groups.

Through programs, science centers give varied audiences access to people who can provide exciting experiences with real stuff and put a tangible, personal face on scientific endeavor. In most cases, these people are staff. But sometimes programs involve local scientists and engineers who come to talk about their work on the weekend. Sometimes staff use special equipment (an electron microscope, for example) in a laboratory-like setting that visitors could not handle alone. They employ a tremendously wide range of techniques—from pocket experiments, to theatrical performances, to contests—to get people of different ages and backgrounds involved. Program staff, often called education staff, conduct programs in nooks on the exhibit floor or in separate rooms in the center. Or they may take programs off site. Wherever they go, they are careful to use materials and inquiry-based methods not found in normal classroom practice.

In the ideal case, exhibits and programs are developed at the same time so that education staff and the exhibit R&D team can interact, sharing knowledge and matching content notions to the most suitable communication technique. Like their exhibit-building colleagues, education staff use feedback from visitors to assess the effectiveness of their various offerings. Fine-tuning a program is somewhat easier than fine-tuning an exhibit since presentations can be adjusted on the fly. This same flexibility makes measuring the overall effectiveness of the program more complicated, though, since presentations need to be checked for "quality control" over time, not just for their potential to inform.[17]

The same few people can't present programs to the public day after day, over and over again, and stay fresh. So education staff engage other people to present repeated programs, sometimes on a part-time or volunteer basis, who may be high school students, college students, science professionals, teachers on contract or on loan to the center, or simply friends of the institution, depending on the requirements of the program. All of these people must be trained and managed in order for them to work effectively in context. In most cases the time and energy required are not trivial.[18]

Science center staff minimize the administrative load by developing activities that can be packaged differently for different audiences and by rotating programs across the year. They vary staff assignments once individuals gain in knowledge and skill. A youngster who gets good at performing a chemistry demonstration on the exhibit floor, for example, may soon give a workshop for families in the chemistry laboratory. Indeed, several science centers have developed mechanisms for nurturing junior staff, particularly females and minorities, and promoting their pursuit of science learning.

Left to themselves, adolescents do not regularly frequent science centers, but as employees or volunteers they serve as lively and energetic floor staff who appeal to visitors of all ages, and they help run the show. Adolescent employees or interns also serve as goodwill ambassadors to their communities, bringing back family and friends. And they benefit over the long term from their involvement with the center and science. Recognizing the potential for mutual gain, a good number of science centers have established means for involving adolescents in their operations. In these efforts, adult leaders directly address the adolescents' distinct developmental needs while also meeting the center's need for effective staff.[19]

In the ideal case, exhibits and programs are developed at the same time.

Floor staff of any age can provide important feedback on visitors' questions to the people who develop exhibits and programs. Communication between developers and floor staff can result in adjustments that enhance the quality of the visitors' experiences. Some science centers involve experienced floor staff early on in the development phase of exhibits or programs, asking them to preview ideas based on their knowledge of visitors' responses.

Science centers offer a wide range of programs that have been described by a number of authors.[20] These programs are geared to three types of audiences: the general public, special target audiences, and the school community. In all cases, programs can be on-site or off-site, temporary or permanent, as the logic of the program requires. *For general audiences, over 80 percent of science centers offer demonstrations or lectures, classes or workshops, and special events* (like soap bubble festivals) on site. Most science centers also target outreach programs to people who are unable to visit the center or who might not come on their own. Also frequently targeted are gifted students and girls and minorities.

Most science centers make a distinct effort to serve the local school community.

Most science centers make a distinct effort to serve the local school community beyond hosting visits by school groups. *Over 80 percent of science centers offer classes or demonstrations for school groups and teacher training.* A key to developing effective programs for students and teachers is to interact regularly with stakeholders in the local education community. The majority of science centers establish steering committees or advisory bodies that include teachers, science supervisors, administrators, parent-teacher organizations, and sometimes the local business/community alliance for science. The steering committee helps determine local science education needs and promotes the center's policies to address them. Appropriately cultivated, the committee can greatly enhance the center's credibility with the school system.

When asked how the science center can best help them, teachers often say they want the center to offer excitement and motivation for children, to provide large-scale or hard-to-maintain apparatus, and to give individual students a chance to pursue their own interests at their own pace. Exhibits and programs developed for general audiences often emphasize precisely these qualities, and so they can readily be made to serve both constituencies. Sometimes schools will also ask for assistance with subject matter studies or teacher training; science centers develop specific programs to fill these needs.

In recent years, science centers have been expanding their service to the school community, leveraging their resources to reach more young people. One result is an impressive array of teacher education programs, described in a 1990 study by Mark St. John.[21] St. John estimated that some 40,000 teachers attended workshops at science centers across the U.S. The workshops ranged from an afternoon spent learning to use a kit of materials to an intensive summer-long institute on interactive teaching methods. Organized jointly with a local university, some programs offered college credit. Teacher education programs at science centers vary in quality, but they are clearly in demand by resource-strapped schools. Science centers have something of value for educators: knowledge, materials, and staff committed to experience-based learning. Teachers who become involved with a science center can help develop programs to put these resources to good use.

A number of start-up science centers have presented programs long before opening their doors in order to demonstrate the kind of service the developing institution will provide. Sometimes these demonstrations have included exhibits, trucked around the region in a van or set up in a school or library, as well as live interpreters. Visiting scientist, school assembly, and pen-pal programs have been used successfully to get things going.

Programs can be administered from an office and taken to locations throughout the community. Sometimes programs can be developed in close collaboration with community groups, who will contribute expertise and resources. Later, when the institution opens, it may choose to continue to conduct off-site programs for targeted audiences. In any event, by starting with programs, a center will have developed knowledge of its community and honed its educational methods before beginning full-scale operations.

N o t e s

1. Frank Oppenheimer described perception as a central theme for the Exploratorium in an article reprinted in Hilde Hein, *The Exploratorium: The Museum as Laboratory*. Robert Sullivan argues that natural history museums should adopt ecology as a primary organizing theme in "Trouble in Paradigms." Michael Robinson makes a similar argument for zoos and other living collections in "Bioscience Education through Bioparks."

2. For example, descriptions of displays can be found in Raymond Bruman and Ron Hipschman, *Exploratorium Cookbooks I, II, III;* descriptions of programs have been published by the Lawrence Hall of Science, in particular the series Great Explorations in Math and Science (GEMS).

3. Mark St. John and Sheila Grinell, *The ASTC Science Center Survey: An Independent Review of Findings.*

4. See, for example, Robert E. Yager, "The Constructivist Learning Model."

5. See, for example, Elsa Feher and Karen Rice, "Development of Scientific Concepts Through the Use of Interactive Exhibits in a Museum."

6. James Bradburne, "Beyond Hands-On: Truth-telling and the Doing of Science."

7. American Association for the Advancement of Science, "Chapter 13: Effective Learning and Teaching." *Science for All Americans.*

8. Roger Miles, "Audiovisuals, a Suitable Case for Treatment"; and Barbara Flagg, "Implementation Formative Evaluation of Earth Over Time Videodisc."

9. Sin's work is unpublished. For a comparison of Sin's work and other research on what constitutes a "successful" exhibit,

see Wendy Pollock and Susan McCormick, eds., *The ASTC Science Center Survey: Exhibits Report.*

10. For description of the role of exhibit developer, see Lisa Falk, "'Not about Stuff, But for Somebody': Michael Spock on the Client-Centered Museum."

11. See Samuel Taylor, *Try It! Improving Exhibits through Formative Evaluation.*

12. For a discussion of reliance on prototyping in exhibit development, see Frank Oppenheimer et al, *Working Prototypes.*

13. For a thought-provoking summary of evaluation techniques, see Mark St. John, "New Metaphors for Carrying Out Evaluations in the Science Museum Setting."

14. See Jeff Kennedy, *User-Friendly: Hands-On Exhibits that Work;* Shab Levy, *Cogs, Cranks and Crates: Guidelines for Hands-On Traveling Exhibitions;* Kathleen McLean, *Planning for People in Museum Exhibitions* (forthcoming). For exhibit graphics, see Beverly Serrell, *Making Exhibit Labels: A Step by Step Guide.*

15. For an approach to developing safety procedures, see Douglas A. Johnston, "The Law of Museum Safety."

16. For further information on accessibility for people with disabilities, see *Barrier-Free in Brief.*

17. See Ross Loomis, "Chapter 7: Using Evaluation to Improve Programs."

18. For aspects of training floor staff, see Caryl Marsh, "Opening the Way for Questions"; and Sheila Grinell and Patricia Curlin, *Using Scientist Volunteers at Museums.*

19. Deborah Edward et al, *Youth Volunteer Programs in Museums.*

20. See Bonnie Pitman-Gelles, *Museums, Magic & Children: Youth Education in Museums;* and Susan McCormick, ed., *The ASTC Science Center Survey: Education Report.*

21. Mark St. John, *First Hand Learning: Teacher Education in Science Museums.*

V I E W P O I N T S

In a classic article, educator David Hawkins describes how children messing around with pendulums can make discoveries and, with the right kind of help, follow their emerging thoughts to gain new insights. His work has influenced science center designers and educators, as has the work of psychologists Lauren Resnick and Michelene Chi, who summarize research in developmental and cognitive psychology about how people learn science.

MESSING ABOUT IN SCIENCE *David Hawkins*

> "Nice? It's the *only* thing," said the Water Rat solemnly, as he leant forward for his stroke. "Believe me, my young friend, there is *nothing*—absolutely nothing—half so much worth doing as simply messing about in boats. Simply messing," he went on dreamily, "messing—about—in—boats—messing—"
>
> Kenneth Grahame, *The Wind in the Willows*

As a college teacher, I have long suspected that my students' difficulties with the intellectual process come not from the complexity of college work itself, but mainly from their home background and the first years of their formal education. A student who cannot seem to understand the workings of the Ptolemaic astronomy, for example, turns out to have no evident acquaintance with the simple and "obvious" relativity of motion, or the simple geometrical relations of light and shadow. Sometimes for these students a style of laboratory work which might be called "Kindergarten Revisited" has dramatically liberated their intellectual powers. Turn on your heel with your head back until you *see* the ceiling—turn the other way—and don't fall over!

In the past two years, working in the Elementary Science Study, I have had the experience, marvelous for a naive college teacher, of studying young children's learning in science. I am now convinced that my earlier suspicions were correct. In writing about these convictions, I must acknowledge the strong influence on me by other staff members in the Study. We came together from a variety of backgrounds—college, high school, and elementary school teachers—and with a variety of dispositions toward science and toward teaching. In the course of trial teaching and of inventing new curricular materials, our shop talks brought us toward some consensus but we still had disagreements. The outline of ideas I wish to present here is my own, therefore, and not that of the group which has so much influenced my thinking. The formulation I want to make is only a beginning. Even if it is right, it leaves many questions unanswered, and therefore much room for further disagreement. In so complex a matter as education, this is as it should be. What I am going to say applies, I believe, to all aspects of elementary education. However, let me stick to science teaching.

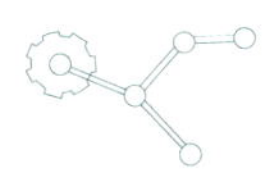

My outline is divided into three patterns or phases of school work in science. These phases are different from each other in the relations they induce between children, materials of study, and teachers. Another way of putting it is that they differ in the way they make a classroom look and sound. My claim is that good science teaching moves from one phase to the other in a pattern which, though it will not follow mechanical rules or ever be twice the same, will evolve according to simple principles. There is no necessary order among these phases, and for this reason, I avoid calling them I, II, and III, and use instead some mnemonic signs which have, perhaps, a certain suggestiveness: ●, ▲, and ■.

● *Phase.* There is a time, much greater in amount than commonly allowed, which should be devoted to free and unguided exploratory work (call it play if you wish; I call it work). Children are given materials and equipment—*things*—and are allowed to construct, test, probe, and experiment without superimposed questions or instructions. I call this ● phase "Messing About," honoring the philosophy of the Water Rat, who absentmindedly ran his boat into the bank, picked himself up, and went on without interrupting the joyous train of thought:

> "—about in boats—or *with* boats. . . In or out of 'em, it doesn't matter. Nothing seems really to matter, that's the charm of it. Whether you get away, or whether you don't; whether you arrive at your destination or whether you reach somewhere else, or whether you never get anywhere at all, you're always busy, and you never do anything in particular: and when you've done it there's always something else to do, and you can do it if you like, but you'd much better not."

In some jargon, this kind of situation is called "unstructured," which is misleading; some doubters call it chaotic, which it need never be. "Unstructured" is misleading because there is always a kind of structure to *what* is presented in a class, as there was to the world of boats and the river, with its rushes and weeds and mud that smelled like plumcake. Structure in this sense is of the utmost importance, depending on the children, the teacher, and the backgrounds of all concerned.

Let me cite an example from my own recent experiences. Simple frames, each designed to support two or three weights on strings, were handed out one morning in a fifth-grade class. There was one such frame for each pair of children. In two earlier trial classes, we had introduced the same equipment with a much more "structured" beginning, demonstrating the striking phenomenon of coupled pendulums and raising questions about it before the laboratory work was allowed to begin. If there was guidance this time, however, it came only from the apparatus—a pendulum is to swing! In starting this way I, for one, naively assumed that a couple of hours of "Messing About" would suffice. After two hours, instead, we allowed two more and, in the end, a stretch of several weeks. In all this time, there was little or no evidence of boredom or confusion. Most of the questions we might have planned for came up unscheduled.

Why did we permit this length of time? First, because in our previous classes we had noticed that things went well when we veered toward "Messing About" and not as well when we held too tight a rein on what we wanted the children to do. It was clear that these children had insufficient acquaintance with the sheer phenomena of pendulum motion and needed to build an apperceptive background, against which a more analytical sort of knowledge could take form and make sense. Second, we allowed things to develop this way because we decided we were getting a new kind of feedback from the children and were eager to see where and by what paths their interests would evolve and carry them. We were rewarded with a higher level of involvement and a much greater diversity of experiments. Our role was only to move from spot to spot, being helpful but never consciously prompting or directing. In spite of—because of!—this lack of direction, these fifth-graders became very familiar with pendulums. They varied the conditions of motion in many ways, exploring differences in length and amplitude, using different sorts of bobs, bobs in clusters, and strings, etc. And have *you* tried the underwater pendulum? They did! There were many sorts of discoveries made, but we let them slip by without much adult resonance, beyond our spontaneous and manifest enjoyment of the phenomena. So discoveries were made, noted, lost, and made again. I think this is why the slightly pontifical phrase "discovery method" bothers me. When learning is at the most fundamental level, as it is here, with all the abstractions of Newtonian mechanics just around the corner, don't rush! When the mind is evolving the abstractions which will lead to physical comprehension, all of us must cross the line between ignorance and insight many times before we truly understand. Little facts, "discoveries" without the growth of insight, are *not* what we should seek to harvest. Such facts are only seedlings and should sometimes be let alone to grow into. . .

I have illustrated the phase of "Messing About" with a constrained and inherently very elegant topic from physics. In other fields, the pattern will be different in detail, but the essential justification is the same. "Messing About" with what can be found in pond water looks much more like the Water Rat's own chosen field of study. Here, the implicit structure is that of nature in a very different mood from what is manifest in the austerities of things like pendular motion or planet orbits. And here, the need for sheer acquaintance with the variety of things and phenomena is more obvious, before one can embark on any of the roads toward the big generalizations or the big open questions of biology. Regardless of differences, there is a generic justification of "Messing About" that I would like, briefly, to touch upon.

This phase is important, above all, because it carries over into school that which is the source of most of what children have already learned, the roots of their moral, intellectual, and esthetic development. If education were defined, for the moment, to include everything that children have learned since birth, everything that has come to them from living in the natural and the human world, then by any sensible measure what has come before age five or six would outweigh all the rest. When we narrow the scope of education to what goes on in schools, we throw out the method of that early and spectacular progress at our peril. We know that five-year-olds are very unequal in their mastery of this or that. We also know that their histories are responsible for most of this inequality, utterly masking the congenital differences except in special cases. This is the immediate fact confronting us as educators in a society committed, morally and now by sheer economic necessity, to universal education.

> **All of us must cross the line between ignorance and insight many times before we truly understand.**

To continue the cultivation of earlier ways of learning, therefore; to find in *school* the good beginnings, the liberating involvements that will make the kindergarten seem a garden to the child and not a dry and frightening desert, this is a need that requires much emphasis on the style of work I have called ●, or "Messing About." Nor does the garden in this sense end with a child's first school year, or his tenth, as though one

could then put away childish things. As time goes on, through a good mixture of this with other phases of work, "Messing About" evolves with the child and thus changes its quality. It becomes a way of working that is no longer childish though it remains always childlike, the kind of self-disciplined probing and exploring that is the essence of creativity.

The variety of the learning—and of inhibition against learning—that children bring from home when school begins is great, even within the limited range of a common culture with common economic background (or, for that matter, within a single family). Admitting this, then if you cast your mind over the whole range of abilities and backgrounds that children bring to kindergarten, you see the folly of standardized and formalized beginnings. We are profoundly ignorant about the subtleties of learning but one principle ought to be asserted dogmatically: That there must be provided some continuity in the content, direction, and style of learning. Good schools begin with what children have *in fact* mastered, probe next to see what *in fact* they are learning, continue with what *in fact* sustains their involvement.

▲ *Phase.* When children are led along a common path, there are always the advanced ones and always the stragglers. Generalized over the years of school routine, this lends apparent support to the still widespread belief in some fixed, inherent levels of "ability," and to the curious notions of "under-" and "over-achievement." Now, if you introduce a topic with a good deal of "Messing About," the variance does not decrease, it increases. From a conventional point of view, this means the situation gets worse, not better. But I say it gets better, not worse. If after such a beginning you pull in the reins and "get down to business," some children have happened to go your way already, and you will believe that you are leading these successfully. Others will have begun,

however, to travel along quite different paths, and you have to tug hard to get them back on to yours. Through the eyes of these children you will see yourself as a dragger, not a leader. We saw this clearly in the pendulum class I referred to; the pendulum being a thing which seems deceptively simple but which raises many questions in no particular order. So the path which each child chooses is his best path.

If you cast your mind over the whole range of abilities and backgrounds that children bring to kindergarten, you see the folly of standardized and formalized beginnings.

The result is obvious, but it took me time to see it. If you once let children evolve their own learning along paths of their choosing, you must see it through and *maintain* the individuality of their work. You cannot begin that way and then say, in effect, "That was only a teaser," thus using your adult authority to devalue what the children themselves, in the meantime, have found most valuable. So if "Messing About" is to be followed by, or evolve into, a stage where work is more externally guided and disciplined, there must be at hand what I call "Multiply Programmed" material; material that contains written and pictorial guidance of some sort for the student, but which is designed for the greatest possible variety of topics, ordering of topics, etc., so that for almost any given way into a subject that a child may evolve on his own, there is material available which he will recognize as helping him farther along that very way. Heroic teachers have sometime done this on their own, but it is obviously one of the places where designers of curriculum materials can be of enormous help, designing those materials with a rich variety of choices for teacher and child, and freeing the teacher from the role of "leader-dragger" along a single preconceived path, giving the teacher encouragement and real logistical help in diversifying the activities of a group. Such material includes good equipment, but above all, it suggests many beginnings, paths from the familiar into the unknown. We did not have this kind of material ready for the pendulum class I spoke about earlier and still do not have it. I intend to work at it and hope others will.

It was a special day in the history of that pendulum class that brought home to me what was needed. My teaching partner was away (I had been the observer, she the teacher). To shift gears for what I saw as a more organized phase of our work, I announced that for a change we were all going to do the same experiment. I said it firmly and the children were, of course, obliging. Yet, I saw the immediate loss of interest in part of the class as soon as my experiment was proposed. It was designed to raise questions about the *length* of a pendulum, when the bob is multiple or odd-shaped. Some had come upon the germ of that question; others had no reason to. As a college teacher I have tricks, and they worked here as well, so the class went well in spite of the unequal readiness to look at "length." We hit common ground with rough blackboard pictures, many pendulums shown hanging from a common support, differing in length and the shape and size of bobs. Which ones will "swing together"? Because their eyes were full of real pendulums, I think, they could *see* those blackboard pictures swinging! A colloquium evolved which harvested the crop of insights that had been sowed and cultivated in previous weeks. I was left with a hollow feeling, nevertheless. It went well where, and only where, the class found common ground. Whereas in "Messing About" all things had gone uniformly well. In staff discussion afterward, it became clear that we had skipped an essential phase of our work, the one I am now calling ▲ phase, or Multiply Programmed.

There is a common opinion, floating about, that a rich diversity of classroom work is possible only when a teacher has small classes. "Maybe *you* can do that; but you ought to try it in my class of 43!" I want to be the last person to belittle the importance of small classes. But in this particular case, the statement ought to be made that in a large class one cannot afford *not* to diversify children's work—or rather *not* to allow children to diversify, as they inevitably will, if given the chance. So-called "ability grouping" is a popular answer today, but is no answer at all to the real questions of motivation. Groups which are lumped as equivalent with respect to the usual measures are just as diverse in their tastes and spontaneous interests as unstratified groups! The complaint that in heterogeneous classes the bright ones are likely to be bored because things go too slow for them ought to be met with another question: Does that mean that the slower students are *not* bored? When children have no autonomy in learning, everyone is likely to be bored. In such situations the overworked teachers have to be "leader-draggers" always, playing the role of Fate in the old Roman proverb: "The Fates lead the willing; the unwilling they drag."

"Messing About" produces the early and indispensable autonomy and diversity. It is good—indispensable—for the opening game but not for the long middle game, where guidance is needed; needed to lead the willing! To illustrate once more from my example of the pendulum, I want to produce a thick set of cards—illustrated cards in a central file, or single sheets in plastic envelopes—to cover the following topics among others:

1. Relations of amplitude and period.
2. Relations of period and weight of bob.
3. How long is a pendulum (odd-shaped bobs)?
4. Coupled pendulums, compound pendulums.

5. The decay of motion (and the idea of half-life).

6. String pendulums and stick pendulums—comparisons.

7. Underwater pendulums.

8. Arms and legs as pendulums (dogs, people, and elephants).

9. Pendulums of other kinds—springs, etc.

10. Bobs that drop sand for patterns and graphs.

11. Pendulum clocks.

12. Historical materials, with bibliography.

13. Cards relating to filmloops available, in class or library.

14. Cross-index cards to other topics, such as falling bodies, inclined planes, etc.

15.—75. Blank cards to be filled in by classes and teachers for others.

This is only an illustration; each area of elementary science will have its own style of Multiply Programmed materials. Of course, the ways of organizing these materials will depend on the subject. There should always be those blank cards, outnumbering the rest.

There is one final warning. Such a file is properly a kind of programming—but it is not the base of rote or merely verbal learning, taking a child little step by little step through the adult maze. Each item is simple, pictorial, and it guides by suggesting further explorations, not by replacing them. The cards are only there to relieve the teacher from a heroic task. And they are only there because there are apparatus, film, library, and raw materials from which to improvise.

■ *Phase.* In the class discussion I referred to, about the meaning of *length* applied to a pendulum, I was reverting back to the college-teacher habit of lecturing; I said it went very well in spite of the lack of Multiply Programmed background, one that would have taken more of the class through more of the basic pendulum topics. It was not, of course, a lecture in the formal sense. It was question-and-answer, with discussion between children as well. But still, I was guiding it and fishing for the good ideas that were ready to be born, and I was telling a few stories, for example, about Galileo. Others could do it better. I was a visitor, and am still only an amateur. I was successful then only because of the long build-up of latent insight, the kind of insight that the Water Rat had stored up from long afternoons of "Messing About" in boats. It was more than he could ever have been told, but it gave him much to tell. This is not all there is to learning, of course; but it is the magical part, and the part most often killed in school. The language is not yet that of the textbook, but with it even a dull-looking textbook can come alive. One boy thinks the length of a pendulum should be measured from the top to what he calls the "center of gravity." If they have not done a lot of work with balance materials, this phase is for most children only the handle of an empty pitcher, or a handle without a pitcher at all. So I did not insist on the term.

Theorizing in a creative sense needs the content of experience and the logic of experimentation to support it.

Incidentally, it is not quite correct physics anyway, as those will discover who work with the stick pendulum. Although different children had specialized differently in the way they worked with pendulums, there were common elements, increasing with time, which would sustain a serious and extended class discussion. It is this pattern of discussion I want to emphasize by calling it a separate, ■ phase. It includes lecturing, formal or informal. In the above situation, we were all quite ready for a short talk about Galileo, and ready to ponder the question whether there was any relation between the way unequal weights fall together and the way they swing together when hanging on strings of the same length. Here we were approaching a question—a rather deep one, not to be disposed of in fifteen minutes—of theory, going from the concrete perceptual to the abstract conceptual. I do not believe that such questions will come alive either through early "Messing About" or through the Multiply Programmed work with guiding questions and instructions. I think they come primarily with discussion, argument, the full colloquium of children and teacher. Theorizing in a creative sense needs the content of experience and the logic of experimentation to support it. But these do not automatically lead to conscious abstract thought. Theory is square! ■

We of the Elementary Science Study are probably identified in the minds of those acquainted with our work (and sometimes perhaps in our own minds) with the advocacy of laboratory work and a free, fairly ● style of work at that. This may be right and justified by the fact that prevailing styles of science teaching are ■ most of the time, much too much of the time. But what we criticize for being too much and too early, we must work to re-admit in its proper place.

I have put ●, ▲, and ■ in that order, but I do not advocate any rigid order; such phases may be mixed in many ways and ordered in many ways. Out of the colloquium comes new "Messing About." Halfway along a programmed path, new

phenomena are accidently observed. In an earlier, more structured class, two girls were trying obediently to reproduce some phenomena of coupled pendulums I had demonstrated. I heard one say, "Ours isn't working right." Of course, pendulums never misbehave; it is not in their nature; they always do what comes naturally, and in this case, they were executing a curious dance of energy transference, promptly christened the "twist." It was a new phenomenon, which I had not seen before, nor had several physicists to whom, in my delight, I later showed it. Needless to say, this led to a good deal of "Messing About," right then and there.

What I have been concerned to say is only that there are, as I see it, three major phases of good science teaching; that no teaching is likely to be optimal which does not mix all three; and that the one most neglected is that which made the Water Rat go dreamy with joy when he talked about it. At a time when the pressures of prestige education are likely to push children to work like hungry laboratory rats in a maze, it is good to remember that their wild, watery cousin, reminiscing about the joys of his life, uttered a profound truth about education.

David Hawkins is distinguished professor of philosophy, emeritus, at the University of Colorado, Boulder.

COGNITIVE PSYCHOLOGY & SCIENCE LEARNING
Lauren B. Resnick and Michelene T. H. Chi

For many years now, Piaget's theory of cognitive development has provided the major framework for science educators seeking guidance from psychological research. In recent years, a new wave of research on human cognition has enriched and expanded our understanding of how people of all ages think and learn. Some of this research examines specific forms of science learning and thinking. Together, Piagetian and Post-Piagetian cognitive research provide a framework for designing informal science education activities that will help people learn science.

PIAGET'S WORK

Piaget's work on children's knowledge of scientific phenomena began early in his career. In the 1920s he conducted a series of studies that educators sometimes overlook in focusing on his much better known "stage theory" of cognitive development. In these studies, Piaget (1930) traced the development of children's ideas about natural and mechanical phenomena, such as winds, clouds, floating and sinking boats, bicycles, and steam engines. Piaget concluded that young children do not believe in the necessity of physical causes, but instead believe that things happen because of the intentions and desires of objects. They believe, for example, that clouds move by themselves; the moon follows people as they walk; the pedals of a bicycle do not have to be attached to the wheel to make it turn. Several decades later, Piaget (1974) focused on children's ideas about force, energy, and transmission of movements. He found that young children attributed movements to forces and energies inside objects and believed that people could move objects without touching them directly.

Children also could not coordinate the effects of several variables. For example, when Piaget asked young children to predict the distance that a ball would travel when shot into motion by a spring-powered plunger, their answers were based on only one dimension of the physical situation, such as the size of the ball. Both the early and later studies showed that children gradually outgrow misconceptions and limitations of these kinds as they grow older. They come to assume and search for physical causes; they combine or systematically "factor out" multiple variables; and they no longer attribute intentions and independent actions to inanimate objects. As the children mature from preschoolers to adolescents, their ideas become progressively more like proper scientific ones. Piaget used these and related findings of mathematical and logical performances of children as the empirical foundation of a complex theory of mental development. That theory has two fundamental aspects—constructivism and logical determinism.

Constructivism expresses the idea that people must build their knowledge for themselves. What a person knows is not a simple reflection of what he or she has seen, heard, or been told. Rather, knowledge is a complex result of a long process of personal construction and interpretation. All knowledge, from apparently simple and obvious ideas about number to the most complex theories of motion, energy, and force, must be built through a laborious process in which new and more powerful ideas gradually challenge and replace earlier ones. This is done by *assimilation,* interpreting new information in terms of already established conceptions, and *accommodation,* adjusting and restructuring initial conceptions to take account of challenging data or ideas proposed by others. Assimilation and accommodation coexist in a nearly continuous interplay in all thinking people, resulting in successive stages of *equilibrium,* or states in which people use current conceptions to interpret and explain experience.

Recent research confirms and elaborates on the basic principle of constructivism. Cognitive scientists describe the role of *schemas*—established, organizing mental conceptions—in understanding and learning about new phenomena. Much research shows what we already know determines what sense we will make of new information—as we read, as we experiment, and as we talk with others. Two major implications of the constructivist principle are that even "low-level" learning involves inferences that go beyond what is actually stated, and subsequent reasoning depends on being able to create "mental models."

Piaget's theory of *logical determinism* claims that as children mature and engage in normal social interchange, they acquire a set of universal logical structures. Although these logical structures are not specific to any domain of knowledge, they determine the kind of reasoning and, therefore, the kind of mental constructions that people will be capable of in any field of science. Piaget claimed that children's prescientific conceptions are primarily a function of the undeveloped logical capacities of children. Their ideas are unscientific because they have not yet developed the general logical structures they need to reason scientifically. According to Piaget, these logical structures develop gradually, in a fixed sequence.

Piaget's three main stages of logical development are well-known: the preoperational, in which children can reason only about physically present relationships; the concrete operational, in which they can imagine certain key actions and effects and thus overcome their dependence on the actual

visible environment; and the formal operational, in which they can generate all logically possible combinations and thus engage in fully scientific "logico-deductive" reasoning. Until this logical development is complete, children cannot appreciate or use key scientific processes (such as controlling variables in experiments or combining vectors) and cannot understand complex scientific concepts (such as physical causality). Piaget used this theory of logical determinism to account for the recurrent finding that scientific misconceptions are resistant to training and yet tend to disappear with cognitive maturity. However, five kinds of evidence show that age-related logical development is not an inflexible prerequisite for understanding advanced scientific concepts.

RECENT RESEARCH

Performance Varies Within a Stage. Many studies show that "stages" of logical development cannot account adequately for the variety of performances seen in children. Children can perform one task (e.g., number conservation) at the level of concrete operations, but still seem preoperational on a number of other conservation tasks (e.g., conservation of weight or mass). Piaget recognized this phenomenon, which he called *decalage,* the French word for time displacement. He explained decalage by proposing that a child can reason operationally when he or she reaches a given level of logical competence, but has to exercise and develop this ability separately for each subject. Piaget thus recognized that specific knowledge plays a role in people's ability to demonstrate logical competence, but he did not admit that knowledge actually helps people to develop their logical competence.

Young Children Reason Logically. Children can also reason operationally well before the Piagetian studies had claimed (Gelman & Baillargeon, 1983). For example, when researchers asked preschool children to decide which of two rows of candies they would like to eat, rather than some abstract (to them) question about whether the number of candies has changed, the children recognized that the number of candies does not increase when the row is spread out. Children recognize that smaller classes of objects (e.g., pine trees) are included in larger classes (e.g., forests) when the wording of the question does not "mislead" them into paying attention to individual objects rather than collections. Children much younger than adolescents can design experiments that systematically vary only one dimension at a time if they know something about the dimensions that are likely to affect an outcome. Furthermore, young children are not necessarily animistic in their judgments of natural and artificial objects, and they reason in terms of sensible physical causality if they have been exposed long enough to a particular physical system to have had time to figure out the causal relations. These findings do not prove that no general cognitive growth occurs, but they do force us to doubt that age-related logical stages strongly limit children's possibilities for learning. They also point to the power of specific knowledge in producing abilities to reason—perhaps even in developing the ability to reason well.

The fact that misconceptions are so persistent suggests strongly that logical development alone cannot ensure good scientific development.

Scientific Misconceptions Persist. Recent research shows that misconceptions about the physical and biological world persist even into adulthood. For example, if college students are asked to draw the trajectory of a ball going off a frictionless cliff at 50 miles per hour, many of them draw trajectories that have either a straight horizontal or a straight vertical component rather than the continuous parabolic curve that would result from the combination of the horizontal and vertical motions. They show the ball moving either horizontally in a straight line off the cliff and only then beginning to fall or curving downward for a while and then dropping straight down. Students explain these drawings by saying that the force causing the horizontal motion eventually dissipates and is "taken over" by gravity, an explanation incompatible with Newtonian inertial theory.

Such misconceptions persist even with excellent formal instruction. For example, many of the students who drew the erroneous trajectories of the ball had taken a formal college course that included Newtonian mechanics. The fact that misconceptions are so persistent suggests strongly that logical development alone cannot ensure good scientific development. On the other hand, the existence of misconceptions shows that people make sense of the world as best they can with the information they have on hand. When their scientific information is incomplete, people develop explanations of their own that may conflict with canonical scientific theories. Once people form such misconceptions, they use them to interpret new information, which makes the misconceptions very resilient. A fascinating example comes from a study by Vosniadou and Brewer (1987), who showed that young children commonly believe that the earth is flat. When they are told that the earth is round, they do not envision it as a round ball, but as a round flat pancake. The children assimilate the new information into established conceptions.

Knowledge Changes Scientific Conceptions. Recent research shows that knowledge, rather than logic, changes scientific conceptions. According to this view, children fail to think scientifically not because they cannot reason logically, but because they have not acquired the main organizing principles for some field of knowledge. Carey (1985) showed, for example, that young children initially decide that an object is alive if the object is similar to humans. Researchers asked four-year-olds if an aardvark, bird, worm, cloud, and tools have babies, sleep, breathe, or have bones. Children did not attribute certain properties that all animals share to animals that are very unlike humans, for example, worms. By age 10, however, children knew enough biology to infer more generally that if something is an animal, it must eat, breathe, and reproduce.

Apparently, children learn about biological properties that define life through a mix of formal instruction and informal reasoning. They do not learn other fundamental scientific concepts as easily. To make consistently correct (Newtonian) predictions about projectile motion, for example, children need to undergo a fundamental ontological shift—from believing that motion is a *change* in state and, therefore, must be explained by some cause (such as an external force or an internally stored impetus) to believing that motion is a state and, therefore, need not be caused. This is a fundamental principle of inertial physics. Recent research (Ranney, 1987) shows that college students do not induce this principle even when they receive lots of empirical feedback on their incorrect predictions and even generalize to similar situations, but show that they have not induced a general principle by not generalizing to more "distant" problems (e.g., situations where there is no gravity).

Knowledge Produces Reasoning Ability. Knowledge may produce or release reasoning ability rather than the other way around. Children of the same age, but with very different familiarity with a content, are differentially able to reason

logically about that content. For example, seven-year-olds who know a lot about dinosaurs infer that a newly presented dinosaur is not a meat eater from the *absence* of certain features. By contrast, less knowledgeable seven-year-olds can only infer that a dinosaur is a meat eater by observing the *presence* of certain features such as sharp teeth (Gobbo & Chi, 1986). Inferring from the absence of a feature is a much more cognitively advanced form of logical reasoning. Children who know more about dinosaurs can reason in that advanced form more than children of the same age who do not have comparable knowledge. Similar studies suggest that how much knowledge a person has strongly affects the quality of reasoning skills.

HOW TO HELP PEOPLE LEARN SCIENCE

Constructivism tells us that people have to build their own scientific knowledge and understanding and that, at each step in science learning, they have to interpret new knowledge in the context of what they already understand. What people currently understand, however, is often scientifically ill-formed and sometimes even fundamentally at odds with proper scientific conceptions. Informal science education—indeed all science education—must address two paradoxes:

- We cannot teach directly, in the sense of putting fully formed knowledge into people's heads, yet it is our charge to help people construct powerful and scientifically correct interpretations of the world.

- We must take into account learners' existing conceptions, yet at the same time help them to alter fundamentally their scientific misconceptions.

We know that misconceptions are grounded largely in lack of knowledge rather than in failures of logic. We can show how the process of assimilation works. But the question of how to produce *accommodation*—fundamental change in current conceptions—remains a difficult one on which cognitive science has not yet made significant progress. Although we cannot offer clear prescriptions for how to help people construct powerful new scientific conceptions, we can point to a few tested principles that can guide educational efforts. These promising possibilities for new approaches to science learning are particularly suited to informal education settings.

Organize Knowledge. A diet of intriguing facts about this or that scientific phenomenon (as is sometimes offered in films and museum displays) is unlikely to produce the powerful knowledge structures that constitute good science understanding. We know that people simply will not remember isolated facts very easily. To remember, people need to incorporate new information into some organizing structure. In any given field, "experts" (people who already have quite a bit of knowledge about the field) can remember significantly more new information than can "novices" (people with little or no knowledge of the field). An expert chess player can remember a chessboard position much better than a novice player, and children who know a lot about dinosaurs will remember more members of a list of dinosaurs than other children. The reason? Experts can assimilate new information into an existing context. They have a coherent body of knowledge, not merely a collection of interesting separate facts. Lots of facts, even on the same topic, do not automatically organize themselves into productive scientific principles. The knowledgeable person's organized concepts can "explain" and "elaborate" the new information to be learned. To use the dinosaur example again, highly knowledgeable children generate many pieces of meaningful information about a new dinosaur that allows them to classify the new dinosaur. Then they can draw upon what they already know about the class rather than learn a list of separate features.

Those who know more, then, learn more. They learn more easily because they use organizing principles to lighten the new learning load. A challenge for educators is to find ways to help beginners in a field quickly acquire the organizing principles that will "bootstrap" them toward the kind of knowledge-based learning capabilities characteristic of people more expert in the field. Educators must find ways to highlight key organizing principles for beginners and invite beginners to use these principles to interpret particular facts and phenomena. These principles, of course, will be specific to each field and subfield of science.

The educator must think not of providing every detail of a scientific structure from some field of science, but of providing the scaffolding on which learners will be able to elaborate.

Making principles explicit for learners probably helps, but well-organized presentations and explicit statements do not guarantee that every learner will use the ideas exactly as they are given. The educator must think not of providing every detail of a scientific structure for some field of science, but of providing the scaffolding on which learners will be able to elaborate their own versions of a basic structure. This notion of providing the scaffolding is particularly appropriate for informal education settings. Informal educators can take advantage of the special media and display opportunities at their disposal, along with the generally high motivation that people bring to informal environments, to highlight a small number of powerful organizing principles for learners.

Give Time and Repeated Learning. Acquiring an effectively organized body of knowledge takes considerable time. People need to be exposed to an idea more than one time, and each time must elaborate on prior information. People need to have the opportunity and must be encouraged to pry deeply into some topics. A special challenge for informal educators is to provide learners with that opportunity, because informal educators cannot impose a formal curriculum with fixed sequences of activities. Series of films or exhibits on different aspects of the same topic or on closely related topics, with carefully planned adjunct activities, can do much to meet the learners' needs. The key is to avoid the temptation to substitute broad "coverage" of topics for depth of treatment.

In this process, providing multiple contexts for any given topic is essential. Exposing learners to a topic more than one time does not guarantee that they will construct organized knowledge. If a topic is repeated but the context is unvaried, a form of *encapsulation* of concepts may occur. Learners will accept new principles or theories but keep them so separate from other conceptions that they have a minor influence on the learners' total mental picture of the world. For example, physics students learn to solve quantitative textbook problems involving Newton's laws correctly, yet behave as if they know nothing about inertial mechanics when asked to think about everyday qualitative physics problems. Educators can counter the effects of encapsulation by showing learners how new principles and concepts relate to many aspects of their existing ideas.

Include Laboratories That Invite and Support Elaboration of Ideas. People learn by personally elaborating on material they are given to study. Students who are passive, whether they are reading a textbook or looking at a lively museum display, do not learn much. Successful learners from physics textbooks, for example, are those who elaborate a lot when they study (Chi et al., in press). In other fields, too, successful students actively relate new, to-be-learned information to their previous knowledge of experiences. Less successful learners,

however, are more likely to reread without elaboration when they study (Bransford et al., 1982). In other words, to learn successfully, people must elaborate so that they can incorporate the new materials into their existing knowledge.

Educators can encourage and support the active, elaborative engagement that characterizes good learning. Educators should include questions and problems for learners when they see films, television, or museum exhibits. Museum educators could pose a problem in an introduction to an exhibit, and visitors could then treat the exhibit as a resource for solving the problem. Educators could have their students discuss the exhibit with other visitors and with the museum staff.

Similar activities for informal learning from film and television are essential. Interactive video based activities might be linked to prepared video materials. Educators can provide laboratories with tools for gathering, reducing, and interpreting data. These ideas are not new to science educators, but cognitive research suggests that they should become increasingly central to informal education efforts.

Promote Discussion. Discussion can help learners to elaborate on their scientific knowledge. People learn better when they discuss and debate ideas with one another. Informal education is ideally suited to capitalize on this natural learning style; people who are captivated by what they see or hear are highly likely to talk among themselves. But just permitting conversation is not enough. Appropriate forms of social interaction need to be "designed into" museum exhibits and into plans for television and other media presentations.

Provide Tools That Help People Construct Mental Models. Much scientific reasoning is more qualitative than quantitative, and scientists and engineers often solve problems as if they

were mentally running models of a system in their heads. Many scientific misconceptions arise because of either an absence of a model or a flawed model—one that lacks key constraints and relationships that scientists know about. Simply confronting people with data that contradict their theories is not enough to cause them to reorganize their conceptions. Learners often reject or ignore data that contravene their current views. One reason for this is that, by themselves, learners often cannot construct a mental model of a system that would behave so as to produce the data. Educators can, however, provide learners with the tools for envisioning physical and natural systems and thus developing new and more appropriate mental models.

> **People learn better when they discuss and debate ideas with one another. Informal education is ideally suited to capitalize on this natural learning style.**

Film and animation, when presented on computers, can provide models of natural or physical systems that learners can actually manipulate and, therefore, appropriate mentally. Representational tools can help students understand theoretical elements and relationships that they cannot observe directly in nature. Such tools make visible the "hidden structure" of the physical and natural world. For example, White and Horowitz (in press) developed a set of computer-based microworlds for force and motion problems that make vectors and force into physically visible and manipulable objects that behave in accordance with Newtonian theory. The microworlds serve as physical embodiments of the elements in a scientific theory, rather than a display of raw natural objects. These representational tools are powerful because they help people to overcome natural limitations in their ability to visualize and manipulate complex interactions and relations. As a result, they allow people to construct strong mental models of scientific theories.

THE CHALLENGE OF INFORMAL EDUCATION

Within the broad spectrum of science education, informal educators face special challenges but also command special opportunities. In informal education, where a required syllabus and examinations cannot be used to keep people studying, the need for depth is probably the biggest single challenge. Informal educators need to provide depth even at the expense of coverage. Informal science educators should invite people to delve deeply into scientific domains and to build organized bodies of knowledge. Only in this way will people be able to remember what they have encountered in informal science settings and use it to develop further knowledge. Repeated exposure and multiple contexts for exploring any topic are also crucial. Otherwise, the new science learned will not penetrate the main body of learners' ideas.

Informal science education requires far more than attractive exhibits or engaging media presentations. Informal educators must invite learners to solve problems and otherwise go beyond the material presented. Knowledgeable persons must be available to help guide novices as they elaborate. Informal educators need to invite and support social interaction explicitly and not just to permit it to happen passively.

The different types of informal education can help learners apply and elaborate their knowledge. Informal settings offer ideal environments for collaborating and debating. Learners can acquire knowledge and ask questions. Media and museum exhibits provide particularly attractive opportunities for using representational and envisioning tools. Educators who supply their students with these kinds of informal learning activities and adapt their teaching methods accordingly will greatly enhance the effectiveness of their work.

. . . .

Lauren B. Resnick is professor of psychology and education and director of the Learning Research and Development Center at the University of Pittsburgh, Pennsylvania. She is founder and editor of Cognition and Instruction, a journal in the field of the cognitive psychology of instruction. Michelene T.H. Chi is associate professor of psychology and a senior scientist at the Learning Research and Development Center at the University of Pittsburgh. She serves on the editorial boards of Cognitive Development and Human Development.

Works on cognitive psychology suggested by Lauren Resnick and Michelene Chi are listed in a separate bibliography on pages 153-154.

Reprinted from Marvin Druger, ed., Science for the Fun of It, 1988, by permission of the National Science Teachers Association.

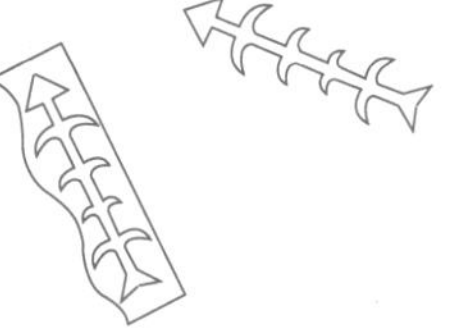

CASE STUDIES

These case studies offer insight into exhibits and programs at four museums. Nils Hornstrup describes the process used by Denmark's new science center, the Eksperimentarium, to select and flesh out content themes, from the initial planning document to refinement of permanent exhibits prompted by feedback from the first year's visitors. The development of a popular traveling exhibition about faces is the subject of Elsa Feher's study, which documents the formation of a heterogeneous design team and explains the decisions made at various stages.

Chuck Howarth explains why the Liberty Science Center used outreach programs to establish a presence in the community years before the building was to open. And Janet Johnson shares her years of experience training staff of the Cranbrook Institute of Science who greet visitors and present programs, explaining both formal and subtle ways to influence the staff's ability to encourage visitor involvement with a science center.

SELECTING THEMES: THE WAY WE DID IT IN DENMARK *Nils Hornstrup*

When you look back, there is always the danger of making things look more ideal, more organized and well thought out than they really were. With this danger in mind, I will try to present the background and the ideas around the overall organization of the Eksperimentarium, Denmark's first science center, which opened in 1991, and how we determined our central themes.

THE FEASIBILITY STUDY

The discussion about creating a science center in Denmark started in 1985 when the Egmont Foundation funded a study on how to implement science centers in the Danish educational system and museum society. I was in charge of carrying out the feasibility study. Very early in this process, the analysis of several European and American science centers led to recognition of the ways that science centers differ from traditional museums. We decided that all of our activities should be selected, built, and presented in a way that every-

one can touch and manipulate, and in this way get real experiences.

But emphasis on phenomena and techniques that can be presented in exciting and interactive ways also contains a weakness, because the communicated information is random and kaleidoscopic. Add to this that the printed instructions and explanations adjacent to the exhibits by nature must be very short to get everyone to use them. The result is that a science center can never try to give the comprehensive information and the perfect picture of a subject. The science center can only be a single—but critical—piece in a broad pattern of possibilities for obtaining basic science knowledge.

Given this position, we realized it would be very important to make sure that single exhibits were presented in a context that is coherent—that makes their meaning clear. We also realized that the very broad target group of visitors at which we are aiming asks for emphasis on basic connections of phenomena to be presented, and that it is essential that the selected exhibits take their starting point in everyday-life experiences. (We later stipulated that these should be everyday experiences as interesting to girls and women as to boys and men.)

We therefore recommended organizing the science center exhibition in a thematic way. The exhibits should be selected and organized in the exhibition hall in so-called theme islands, where working with one exhibit, by concepts and by content, complements working with another related exhibit. The feasibility report also recommended the use of explainers to motivate dialogue about science matters, and program activities to supplement and increase the value of the hands-on exhibits.

We also identified a role for the science center as a supplementary resource for science teachers, helping them organize their own teaching. This role must influence the overall selection of exhibits and themes.

It is essential that the selected exhibits take their starting point in everyday-life experiences.

THE PILOT EXHIBITION

To investigate visitors' interests, to test development and production techniques, and to facilitate the fundraising process, we started by building and running a pilot exhibition, *Your Body.* The exhibition included 60 hands-on exhibits dealing with the physics, chemistry, and biology of the human body. The selection of exhibits was made from a long list of ideas gathered from many science centers around the world,

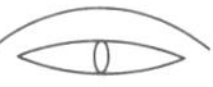

and from discussions with Danish experts from universities, hospitals, and private companies. During the three months of display in 1988, we also tested the use of explainers for small demonstrations, the running of a small café, and a gift shop. As program activities, we included 11 Saturdays of demonstration lectures in the café entitled "Dialogue with the Expert." We also organized a school reservation arrangement to measure the interest from the education system.

Nearly 72,000 people, including about 20,000 school children, visited the exhibition, and from many interviews and discussions we learned that our focus on the human body, the senses, and health was very important and popular. It therefore became essential that the permanent science center should place people at its center.

THE PERMANENT EXHIBITION

By the beginning of 1989, the fundraising campaign had been so successful that the board gave the starting signal for the establishment of a permanent science center. The platform for planning and development was the recommendations of the feasibility study, the experience gained from the pilot exhibition project, and the skills of a small project staff. We started up by presenting our ideas to a group of Danish advisors and experts covering different levels of education, research, museums, and entertainment. In two two-day sessions, we discussed the educational value of science center activities in relation to the different levels of the Danish educational system, the possibilities of collaboration with university experts, and possible cooperation with private companies and research institutions. As a result of these, we made a tentative plan for the overall content of the permanent exhibition, stating essential themes and several ideas for specific exhibits.

The next step in the process was the selection and composition of the development staff. Besides having a certain amount of pedagogical experience in the group, the tentative list of themes and exhibits indicated an academic staff covering physics, chemistry, biology, communication, and nutrition and health. This group was supplemented by an industrial designer, an exhibition architect, and a production team of skilled craftsmen in graphics, woodworking, metalworking, and electronics.

At the end of every period we were left a little frustrated, but with better plans and better ideas.

In May 1989, we started a training and qualification program for this group. This was arranged in a deliberate attempt to get the group to function as a real team and to secure among the staff members the highest level of information about every aspect of the project. During one month we discussed the science center philosophy, our mission statement, and our objectives. We presented information and ideas from other science center exhibitions through slides and videos, discussed different ways of building exhibits, and tried to update the group with the experiences from our pilot project. At the end of the training period, the whole group traveled to Finland to visit the newest science center in Scandinavia.

During the next few months, the scientific staff used nearly all their time conducting primary research on selected exhibits. Every exhibit idea was described in a short synopsis, including the way to use it, the science to learn from it, the way to build it, equipment involved, budget for the its production, persons of reference, and staff member responsible. The production team worked on building a few selected, obvious exhibits and renovating exhibits from the pilot project, and the designer and exhibition architect worked on overall set design and exhibit design manuals. During this period, our plans for the thematic organization of the exhibition and the selection of exhibits for every theme island were revised, as we struggled for agreement among ourselves about how far to go with particular ideas.

EXTERNAL CONSULTANTS

To bring in know-how from experienced institutions abroad, and as a way of further training and qualifying the staff, we hired external consultants during the planning period to critique our plans. In three short sessions we had the pleasure of first meeting Dan Goldwater, former exhibit director of the Franklin Institute Science Museum, Philadelphia; next, Pamela Rogow and Jeffrey Bernstein from R+B Design+Development in Los Angeles; and finally, Steven Pizzey from Science Projects, London. This process proved to be extremely valuable to our group because of the very different views and opinions they brought to us. Every time we had to promote our plans and ideas, and through discussions from very different angles, we received know-how on exhibit content and production, exhibit design, set design and decoration of the actual exhibition hall, graphics, exhibit-building methods, and the selection of materials. At the end of every period we were left a little frustrated, but with better plans and better ideas.

THE PRODUCTION PERIOD

At the end of September 1989, the final plans for the permanent exhibition were settled. The exhibition hall was divided into three equally-sized areas dealing with Man, Nature, and The Interaction between Man and Nature. A corner of the exhibition hall was to be fitted out with a 150-seat grandstand for demonstrations and special events, and a smaller closed area dedicated to scheduled visitor activities, called the "Focus on ..." workshop. The three main themes were subdivided into 15 theme islands, each having 10-20 exhibits. The titles of the theme islands were:

MAN ....
The Body in Action	Your Senses
The Living Cell	The Brain and Senses
Food and Drink	

THE INTERACTION BETWEEN
MAN AND NATURE ....
Communication	Our Environment
Unique Materials	We Use Energy
Mathematics	

NATURE ....
Atoms and Radiation	Look at the Light
Air and Water	Sound and Vibrations
The Laboratory	

The final plans for the theme islands were determined with as few loose ends as possible, but every theme had at least 20 percent more titles than expected to be produced by the scheduled opening. This philosophy proved to be valuable, because good ideas often proved to be impossible very late in the process.

Responsibility for the production of the selected exhibits was shared by the academic staff members according to their skills and interest, supplemented with a team of part-time exhibit researchers in areas where we did not possess sufficient knowledge among the permanent staff. With the very flat organization of the Eksperimentarium, it has been possible to maintain short decision-making processes and a very high level of shared information throughout the whole development and production period.

In total, we developed and built 165 new exhibits and renovated 60 exhibits from the pilot exhibition in the 15-month production period. The number of exhibits at the opening was 225. In the capital budget there was money for another 25 exhibits, but we chose to let the selection of these await the public's response.

FEEDBACK FROM THE VISITORS

Since opening in January 1991, we have been obtaining some raw data and informal feedback from visitors on how they use the exhibits and other facilities. Very early we discovered that we should put more emphasis on children ages 5 to 10. Therefore, the remaining portion of the capital budget was used to develop and produce a new theme island called The Children's Water Puddle. This is an 800-square-foot area with about 11 tons of water circulating in pools in three levels. The area contains water bridges, canals and rivers, and dams and floodgates; children can manipulate pumps, waterwheels, fountains, and faucets. Since opening in December 1991, The Children's Water Puddle has been very popular.

Our staff have observed that visitors don't always perceive the "invisible" borders between one theme island and the next. (We deliberately made them invisible, thinking that we might change exhibit placement.) Many visitors have asked for help in finding their way around. We intend to provide it, first by producing brochures that say, "If you have one hour and are interested in light, see exhibit numbers so and so." We use a lot of effort to keep everything in the exhibition hall functioning smoothly; simultaneously, the development of new programs, written material, and demonstrations to sustain and deepen visitors' interests in our exhibits and themes now have the highest priority.

But we keep on discussing the exhibits and themes. It turns

out that some themes, such as the environment and technology, are harder for visitors to grasp than others. Each of our staff members has had to acknowledge that "a fine and very essential exhibit dealing with an important scientific subject and promoted by myself" is of practically no interest to most visitors. But each of us has had successes as well, and we have confidence in our organizational method. We know that our well-developed, unique thematic plan needs to undergo a long series of changes, revisions, and improvements. This continued renewal will challenge us to maintain our creativity and enthusiasm.

Nils Hornstrup was the project director of the first Eksperimentarium pilot exhibition and since 1989, has been director of development, responsible for the selection, development, and building of hands-on exhibitions. A former professor of experimental physics, he has been working with educational television, science communication, and in-service training for teachers for more than 20 years.

ABOUT FACES: CONCEPTUALIZING AN INTERACTIVE EXHIBITION *Elsa Feher*

A few years ago, a group of eight museums formed a collaborative with the purpose of developing and sharing high-quality interactive exhibitions. Our museum's contribution to this collaborative was *About Faces,* a 1,500-square-foot exhibition that opened in 1988 in San Diego, is still traveling in the United States and Canada, and may soon travel in Europe and South America.

The questions I am addressing in this case study are ones I am often asked: How did you think of choosing the subject of faces? Once you chose the topic, how did you go about developing it? How did you decide what to include, what to leave out, and how to make it a coherent whole? What is it about this process that made the exhibition so successful?

GETTING STARTED

Faces is a somewhat unconventional topic in that it is not a traditional scientific or technical subject such as, say, energy or insects or robotics. The topic suggested itself to us because of the unprecedented popularity of one exhibit we had put on the science center floor. The exhibit, called Sym-ulations, is a computer-based device that captures the image of a visitor's face; users can operate on this image to see what their faces would look like if they were perfectly symmetrical, made up of features from the right side only or of features from the left side only. Visitors can also use their own faces to produce one-eyed, fork-tongued creatures. The success of Sym-ulations is clearly due, first and foremost, to the fact that it draws on people's delight with their own faces (mirror-based exhibits are also very popular). In addition, the exhibit is humorous (it is amusing to produce strange creatures) and surprising (most people are not aware that the right and left sides of their faces are really quite different, and the exhibit allows a fast comparison between the two). The idea for such an exhibit had been floating around the science center for some time, but the chance to develop it came when we met Ed Tannenbaum, a computer graphics artist based in the San Francisco area, who had the expertise and the interest to take on the project.

The most popular exhibit offers surprise and humor, and allows visitors to explore the limits of their human identity.

Through Sym-ulations we learnt a great deal about the use of faces in exhibitry. But to develop a whole exhibition on the subject, we needed to find out what was known about faces that was intriguing, that mattered, and that could be told in an interactive manner. Terry Landau, a Los Angeles-based writer and producer of documentary films, took on the task of researching the topic and preparing an overview that offered an initial treatment of the subject. Landau's story started with the evolution of the face, from gills to jaws to the 14 bones and 80 muscles that support and animate the human face. The narrative took us through physiognomy (the reading of character in the individual's features) to the idea of the face as a pattern, an entity that is recognized as a whole. The story went on to the probable evolutionary process of using facial expressions for communication, Darwin's interpretation that facial expressions are useful for survival, and the notion that some expressions are universal, that is to say, that they express the same emotion in any culture in which they are used.

Landau's overview identified individuals who had done work we might find useful. One of these was Nancy Burson, a New York artist who produced photographs of faces that are interesting composites (such as a woman and a cat, or Nixon and Mao). Burson and her husband Kramlich also developed a computerized system that allowed them to produce a reasonable likeness of people's faces in their old age. We were able

to interest Burson and Kramlich in producing an exhibit for *About Faces* that would use their system and incorporate Ed Tannenbaum's technique for using the visitor's face. The resulting exhibit, Mix and Blend, allows visitors to exchange or blend their features with faces chosen from a data bank of well-known personalities. The data bank includes a variety of ethnic origins as well as fictional faces. In this way the visitors can try on, say, John F. Kennedy's eyes and Oprah Winfrey's mouth. Visitors can also produce, for example, a visitor-Frankenstein composite. This is probably the most popular exhibit in *About Faces*; it has many of the same characteristics as Sym-ulations: It offers surprise and humor, and allows visitors to explore the limits of their human identity.

The Landau overview also directed us to Paul Ekman, professor of psychology at the University of California in San Francisco, expert in the link between facial expressions and emotions. A small group of us traveled to meet with Ekman. From Ekman we learnt how to produce the six universal expressions and emotions (anger, fear, surprise, happiness, sadness, disgust) using the appropriate muscles in our face. He showed us how to detect emotions of extremely short duration (microexpressions) that occur when people try to control their emotions and conceal how they feel. He taught us that, at a distance, some expressions are easier to recognize than others, and also that some expressions are more difficult to mask than others.

After meeting with Ekman it became clear that we would focus the exhibition on the face as an instrument of communication. We now knew that there were enough interesting, dramatic effects that had the potential to become interactive exhibits. Specifically, the exhibition would show that there are messages and meanings in both features and expressions. "Features that convey identity, expressions that communicate

emotions" became the byline for *About Faces*, and our point of view for the exhibition. The scientific content—the effects and phenomena—displayed at the individual exhibits would present and support the exhibition's point of view. The visitors, playing and working with these exhibits, using their faces as objects of study, would find out how identity and emotion are signaled, and so internalize the exhibition's message.

EXPANDING IDEAS

It was November. We had 18 months before opening date. Our in-house development staff consisted of Paul Avery, designer, builder, and sorcerer (provider of the magic touch); and myself, functioning as conceptualizer and project leader. Ed Tannenbaum signed on as consultant in charge of coordinating the effort involving computer graphics exhibits.

Some ideas fell by the wayside, others metamorphosed, and a few saw the light of day as conceived.

We had arrived at a set of concepts that we felt were important, and a corresponding set of ideas for exhibits that needed to be examined for feasibility. To help us develop these ideas into creative exhibits, we set up a brainstorming-with-show-and-tell meeting to which were invited an assortment of individuals with technical, artistic, and academic backgrounds, all of whom had developed something that seemed useful or promising. Since the geographical mean for the participants' dwellings fell somewhere in the Bay Area (some people came from Oregon, others from Los Angeles), we asked the Exploratorium if we could set up the meeting at their place and they graciously agreed. It was a grand day, with people coming and going, showing materials they had produced, playing about with ideas and possibilities. The result was an expanded horizon, a plethora of leads to follow, and a handful of new individuals with whom to work.

We embarked on several projects. A San Francisco sculptor was to produce masks that portrayed the six universal expressions. The ultimate purpose was to make half masks (upper and lower portions of the face) that the museum visitor could put on to practice making faces that complement, override, or blend with the emotion expressed in the mask. An Oregon professor was going to implement for visitor use an ingenious technique, which uses facial features as coordinates to represent data with many variables. A Hollywood special effects specialist was going to produce a rubber mask with muscle-like attachments that could be controlled manually by the visitor to produce different expressions. A San Diego police consultant started development of a digitized bank of facial features, based on the transparencies from the identity kits that are used by the police to enable eye witnesses to reconstruct the faces of criminals. Ed Tannenbaum was also developing exhibits, one dealing with microexpressions, another with recognition of identities and of emotions.

EXHIBITS THAT PROSPER AND EXHIBITS THAT DON'T

The San Francisco sculptor could not manage the steps involved in going from detailed drawings to modeling clay to plaster of Paris molds to a fiberglass product. But we had the good luck to find a San Diego artist who took over the project and did a wonderful job with it. The project that was going to use facial features to represent data was abandoned because we could not come up with interesting enough data to motivate visitors to use the exhibit. The visitor-activated

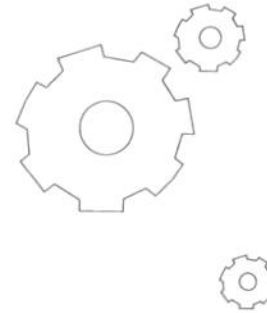

emoting rubber mask project died when the Hollywood specialist investigated sources of rubber and reported that the material used for King Kong and other reconstructions would not last long and would need to be replaced often. We were a few thousand dollars into the development of the computerized identity kit when we heard that a software package had just appeared on the market that did everything we were trying to develop. So our development project was dropped; the developer of the commercial software created a visitor interface per our specifications, and the exhibit was on the floor in a jiffy.

So it is that some ideas fell by the wayside (mostly for technical difficulties or human inadequacies); others metamorphosed; and a few saw the light of the day, as prototypes, pretty much in the form in which they were conceived. Ultimately, all exhibits underwent some modifications during their evolution from prototypical to final form as a consequence of the process of evaluation with the public.

ACHIEVING CLOSURE

The development process can be so exciting that one sometimes wishes for infinite funds and time to just keep going with the great ideas and new possibilities that keep cropping up. The reality, however, is that at a certain point in time one needs to take a view of the whole, make the necessary adjustments, stop development, and go into production. So, six months before opening we paused to ask ourselves whether we had included all that was necessary for the treatment of the topic, if there were conceptual gaps that still needed to be filled, if there were exhibits that did not merit inclusion. We had close to 20 different activities (I don't call them exhibits because some exhibits used the same hardware to present multiple activities). Were these to constitute the final show? It was difficult for us to get enough distance from

the details of the project to trust our own answer to the question. So we had another meeting with our core consultants, this time including Shab Levy, an independent exhibit producer with many years of experience in the field.

At this meeting we decided to incorporate three more exhibits that were easy to add without substantially altering the conceptual design plan that was already in existence. One of the exhibits, Muscle Mirror, responded to a gap in the overall presentation: In the text accompanying several of the exhibits, we referred to the facial muscles that are involved in producing the six universal expressions. Yet these muscles were never actually shown. Another exhibit was added to create a better balance between the number and impact of the units about features and those about expressions. It dealt with our ability to recognize a face when some of its features are rotated, and was simple to include since it could use the same hardware, and even the same title, as another existing exhibit. The third exhibit we added was a simple set of perpendicular mirrors with which the visitors can create the same effects that are seen in Sym-ulations simply by positioning their face along the edge of one mirror and looking into the other. This inclusion was in line with our intent to provide, whenever possible, a low-tech version of the high-tech exhibits.

PAYOFFS AND LAYOUTS

At each exhibit there is a payoff for the visitor: something intriguing to explore, a puzzle to solve, a challenge to fulfill, a surprise, a laugh. There is also a larger payoff if the exhibition becomes a coherent whole rather than a series of unrelated experiences. This is partially accomplished through the physical layout of the exhibition on the museum floor.

Although, as is usual in the initial planning, the *About Faces* overview had been done in the form of a storyline, the actual floor plan does not dictate a linear, sequential flow. In this sense, *About Faces* is not like a book or a movie or a school lesson. I would rather liken it to a happening, those theatrical audience-participation events that were popular in the sixties, staged yet seemingly spontaneous, dependent on the context so they never took precisely the same form.

A visit to *About Faces* does not have prescribed beginning and end points for reasons both practical and philosophical. From the practical standpoint, a prescribed route in our museum would cause traffic bottlenecks. From a philosophical standpoint, we want to encourage a visit that is free from strictures and constraints, where visitors can stroll and sample, choose to stay at an exhibit or move on. An organizing scheme that is not linear can deliver a powerful message if the exhibits are grouped in a contextual manner, placed according to a floor plan that is a network of related ideas, a concept map of sorts that allows the visitor to explore the whole in a variety of different orders, all of which make sense.

Let me illustrate this. *About Faces* has a central core of five computerized exhibits. Assume a visitor starts at one of these, Discernibility, by exploring pattern recognition. At the adjoining exhibit, Recognitions, the visitor is challenged to pick out a known face in a group of unfamiliar ones. Away from the core, behind Discernibility, is Recreations, where the visitor can select from hundreds of different features in order to reconstruct a remembered face. In this way, each exhibit is surrounded by others with slightly overlapping ideas, so the visitor moves in an environment that provides conceptual reinforcement and gradual change from one topic to another.

FINAL MUSINGS

I think there are two characteristics of the process that we lived through in developing *About Faces* that bear pointing out. The first is very specific to our own circumstances, but undoubtedly other museum staff find themselves in similar circumstances. We started out with very restricted in-house capabilities, and hardly any staff. The development team was not in place at the outset, but was put together as part of the exhibition development process. The bulk of the team consisted of outside consultants whose capabilities were demonstrably what was needed for the project. However, the overall vision, the direction needed in the task of turning one consultant's knowledge into another consultant's three-dimensional rendition, remained at all times the responsibility of the museum staff.

The second characteristic is quite general, probably not only for developing exhibitions but for other creative processes as well. It has to do with the rhythms of the process. Starting from the nucleus of an idea, we first went through a flowering or blooming phase, very tuned to the outside world, reading, listening, collecting suggestions. Then came a feasibility study phase, finding the tools and people to implement ideas. And finally the closure phase, when we examined what we had, tightened and modified its content, and said "This will be it."

Elsa Feher is director of the Reuben H. Fleet Science Center and professor of natural science at San Diego State University. She received her Ph.D. in physics from Columbia University, and worked in experimental solid state research for several years at the University of California, San Diego. She has been actively involved in the science education of teachers, in the development of a highly-interactive college-level science curriculum, and in research on cognitive issues.

RIGHT FROM THE START: THE ROLE OF OUTREACH PROGRAMS IN LAUNCHING A SCIENCE CENTER

Charles H. Howarth, Jr.

If you're building a science center, where better to begin than with outreach programming in your community?

I know what you're thinking: "Shouldn't I wait until we're open? I mean, we've got so many other things to get done...and I've got my hands full already trying to raise enough money to get my project off the ground..."

At Liberty Science Center, we thought the question should be whether there is any reason to wait, rather than whether there is any reason to start. After all, the entire point of starting a science center is to support and promote science education in your community (or at least that is our mission at Liberty Science Center; if your mission is different—say, to promote tourism—than perhaps outreach isn't for you). And we found several other good reasons to start our programming early: to establish credibility; to build a staff; and to stay connected to our roots as educators. Let's look at each of these.

ESTABLISHING CREDIBILITY

To be or not to be: That is the question.

Well, not for you, of course. You've already made up your mind that you're going to build your science center, come what may. But other influential members of the community undoubtedly don't share your dream. Not that they oppose you—they just haven't thought about it at all. Some have no idea what a science center is; others hold to an outmoded notion that science centers are mostly a light family entertainment rather than the serious educators we have become. As Roger Nichols, late director of the Boston Museum of Science, once put it, too many community leaders still see museums as nothing more than "a nice place to take your kids on a rainy Sunday."

Your community doesn't know about all of that yet, and one of your first and most important tasks is to educate them; how better to do it than by actively demonstrating your commitment to learning and your ability to deliver through outreach programming?

TESTING YOUR PROGRAMS AND BUILDING A STAFF

The play's the thing!

Nobody opens a Broadway play on Broadway—they go on the road first, where the bugs can be worked out beyond the glare of publicity that comes with opening night.

You face the same problem with your science center. The building itself can't very well go on the road, but some of your exhibits and programs can. At Liberty Science Center, for

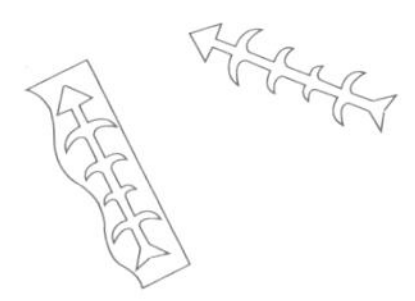

example, we have sent traveling exhibits around New Jersey and are currently field-testing our first science demonstration in front of school groups at the nearby Newark Museum. We have been able to test ideas, critique one another's presentations, train new demonstrators, and get a reaction from a real live audience—all without the pressure we'll face at our opening.

MAINTAINING YOUR SANITY

Tomorrow and tomorrow and tomorrow creeps in this petty pace from day to day...

With persistence, hard work, and a bit of luck, you will get your science center off the ground. But nobody said it was going to be easy, and don't expect success overnight. Liberty Science Center's gestation from earliest discussions to opening spanned 14 years, not unusual for the launch of a new museum. So patience is the word.

This raises the question of what you are going to do over that long period to keep yourself and your staff focused on your mission and the people you hope to serve. Business planning, market research, architecture, and building a board are all essential tasks, but none of them are very satisfying for those of us who got into museums in the first place for the pleasure of working with the public.

For this purpose, outreach can be a lifesaver—a daily reminder of what you are all about, a point of contact with your audience, a chance to reinvigorate yourself with the joy that comes from working in informal education. In those down times that you will inevitably encounter, when funding seems in doubt or the city council just turned down your proposal or the architect explains why you can't afford the building you really want, the smile on a child's face in one of your outreach programs will provide the sense of purpose and fulfillment that you will need to keep going.

CAN WE AFFORD IT?

Sure, outreach programs require resources—not only money, but valuable time that you might feel could be better spent elsewhere. At Liberty Science Center, though, neither time nor money has proved to be an overwhelming obstacle.

First, financial support: As your capital campaign proceeds, many of the foundations you approach—and some corporate sponsors as well—will be unable to provide grants for bricks and mortar. As a result, grants will be available that can only be spent for programming. One major New Jersey foundation actually approached me and requested a proposal for outreach programming. We've had several donors since who have similarly restricted their support.

> **Outreach can be a lifesaver—a daily reminder of what you are about, a point of contact with your audience, a chance to reinvigorate yourself.**

In addition, many outreach programs will partly or even wholly pay for themselves through participant fees. Be careful, however, about programs that others say are "self supporting." Museums keep their books in different ways, and you need to know what expenses are included: materials and supplies? salaries? administrative expenses? As a start-up organization, you will need to find some way to cover all of these expenses, and participant fees alone usually won't be sufficient. As a result, it is essential to have back-up financing during the pilot phase at least.

As for time—if you think of outreach as the heart of what you're all about rather than a distraction, time needn't be a problem. Insofar as outreach helps with fundraising and community support, the effort you put into programming can provide a big boost toward getting the rest of your project off the ground.

EXHIBITS OR PROGRAMS?

Many of the goals listed here can be accomplished with traveling exhibits, but exhibits offer some difficulties. Where will the exhibit be on display? If at other museums in your region, you may find (as we did) that the public associates the exhibit with the host institution rather than with yours. Who will provide maintenance? For an interactive exhibit, this is a big problem unless the host institution has staff experienced in interactive displays. How will the exhibit be shipped, set up, and knocked down?

In addition, your first exhibit needs to represent you well, which means spending money to do it right. And it needs to be large enough to suggest the scope and importance of your project. Our experience has been that programs are easier to do well, easier to manage, and much less expensive to develop and operate than are traveling exhibits.

On the other hand, exhibits will eventually be a major part of your program, and there is no better way to gain experience and test ideas than by actually trying things out with a real audience. So you may want to put some exhibits on the road despite the obstacles.

CHOOSING THE RIGHT PROGRAM

It starts with mission. At Liberty Science Center, we spent six months defining who we are and what we want to accom-

plish. We also surveyed the needs of our community, interviewing educators, government officials, and business and community leaders to get thoughts on where we might fit in.

Some of our findings:

- In-service training for teachers is a critical problem, especially at the elementary level.

- Liberty Science Center is at the heart of the most densely populated and ethnically diverse region of the United States, and urban and minority education is a priority concern throughout our region.

- Career awareness is a priority, not only among educators, but among the corporate scientists who conceived and initiated the center.

- We found strong concern over the need to offer research opportunities for exceptional students.

In response, we selected three audiences for our first outreach programs: elementary teachers, gifted and talented high school students, and urban minority students. In addition to supporting our mission and meeting the needs of our community, we wanted our first programs to help build our image. Since first impressions are hard to change, we thought it essential to choose programs that represent what we want to be: hands-on, highly educational, motivating, fun, and of the highest quality.

LOOKING FOR MODELS

With these goals in hand, we set out to find the best outreach programs already in place at other museums. Not that our programs are exact copies of those elsewhere: Usually it is necessary to make modifications to reflect local conditions, budgets, and the skills and interests of our staff. On the other hand, we felt no need to reinvent the wheel. Programs that have succeeded elsewhere are likely to work for us as well. We examined programs for gifted students at the Maryland Science Center, Boston Museum of Science, and New York Hall of Science; and programs for minority children at the Science Museum of Connecticut, The Carnegie Science Center, and The Brooklyn Children's Museum.

OUTREACH TODAY AT LIBERTY SCIENCE CENTER

Today, four years later and one year before opening, Liberty Science Center has in place several major programs that reach nearly 10,000 participants each year.

Some of the highlights:

- Science-by-Mail, a program designed and franchised by the Boston Museum of Science, pairs elementary students and scientists as pen pals. Our chapter has nearly 1,000 groups participating and is nearly self-supporting from registration fees.

- Skygazer, a program that trains elementary teachers to use a Starlab portable planetarium and then lends the equipment for use back in the classroom. This program is modeled on one at the New York Hall of Science.

- Partners in Science, a summer program that places gifted high school students into corporate and academic laboratories for an intensive eight-week research experience under the guidance of a scientist-mentor—an idea we borrowed from the New York Academy of Sciences and the Museum of Science in Boston.

- IDEAS, a program for youth leaders in Newark and nearby communities that provides training in hands-on science

for use with the families of the young people they serve. The program is patterned after Family Science, a national program developed by Northwest EQUALS of Portland, Oregon.

- Science Design Partnership, a program that encourages teachers to engage their classes in hands-on activities that can be developed into classroom exhibits, and then to share the results with us. At our opening, we will display the best of their efforts, and may even develop some of their ideas into full-fledged exhibits of our own. Eight hundred classrooms are participating.

As a result of these programs, Liberty Science Center is already recognized around New Jersey as a leader in science education. We have a large and growing pool of friends who are eagerly awaiting our opening, an extensive mailing list, and a staff of six full-time educators who are experienced and ready to go. We have organizations approaching us to develop joint grant proposals. Our corporate sponsors are pleased to see their gifts producing such visible results, and are providing additional support as a result. We are poised and ready for our opening. And, by no means least important, we have been able to begin to fulfill our mission by offering high-quality education services to thousands of children across the region.

Charles Howarth is president of the Liberty Science Center, a science museum scheduled to open in October 1992 in Liberty State Park, New Jersey. Mr. Howarth spent 11 years at Boston's Museum of Science as educator, exhibits planner, administrator, and associate director. Before entering the museum field, he taught biology and marine science in New Hampshire and the Virgin Islands.

TRAINING IS NOT TRIVIAL! *Janet Johnson*

When your first exhibits are finished, and your doors are open, it's tempting to think your job is done. Just let the people come and enjoy themselves. The exhibits will be so engaging that you won't need people to explain, to assist, to do programs, to answer questions. The exhibits will take care of everything.

Wrong! Your job has just begun. A major message of any museum is one conveyed through its people—its admission staff, explainers, telephone answerers, custodial staff, volunteers—virtually everyone who is part of your museum family. They all need to have a grounding in your museum's culture, to know where it has been and where it is going.

There are some logical steps that can lead to an effective training model for your museum. Here are some tips from Cranbrook Institute of Science's experience.

MISSION STATEMENT

Use your mission statement as a basic foundation for training. This is the first level of common understandings—what you're all about. This succinct statement of your institution's purpose needs to be readily available to all staff and volunteers, to be used as a coalescing focus for each person's efforts. Print it often in your newsletter, in brochures—even on a plasticized card for staff to carry with them. Show that it's important!

MESSAGE STATEMENT

Develop a very brief and energizing message statement that represents the "feel" or the "spirit" of your institution. Here's an example of a message statement: "Cranbrook Institute of

Science will excite the imagination of visitors to the elegance and power of science as a process for understanding our world and our future." This type of statement has an emotional message and can be used as a guide for exhibits and programming, not just training.

A major message of any museum is one conveyed through its people.

LEARNING MODEL

Structure a learning model based on the mission and message statements. Identify key words that demonstrate how you want visitors to learn in your museum. This might include words like inquiry-based, experimental learning, discovery—whatever you determine best fits your institution. Use these as a guidepost for your teaching methods, the practical methodology that should be used with visitors.

GETTING STARTED IN A TRAINING PROGRAM

Once the philosophical base is in place, you are ready to develop a training program. A new museum is in a great position to start everyone off approximately together—a real advantage. At Cranbrook we have long-tenured employees and volunteers with individual perspectives on what the institution used to be, which is not the same as what it is now or what it will be in the future. A large amount of retraining must be done in order to move forward with a new vision. Celebrate your fresh start with a thoughtful, documented training program that captures the excitement of a new enterprise and focuses everyone's energy in the same direction!

TRAINING HANDBOOK

While the main thrust of a training program should not be merely a formal manual, such a booklet ensures that common information is available in a single place and can be reviewed at leisure.

Although you may choose a complex manual or a simple one, it is probably better to start out with essentials in a folder or notebook to which pages can be easily added. If it gets too big, it doesn't get read and it also becomes so costly to reproduce that you become reluctant to freely use it for volunteers as well as for staff.

Sample table of contents:

- General information—letter from the director, history of the museum/center, logo and explanation of it, mission statement, message statement

- Building information—floor plan, square footage, map of area and grounds

- Organization chart—personnel with names, phone numbers, and position titles, reporting relationships, board of trustees

- Educational information—learning model, programs available, role of explainers

- Policies and procedures—personnel policies, safety and security procedures

- Volunteers—opportunities, responsibilities, benefits

- Answers to common questions

TRAINING AS A FACTOR IN STAFF RECRUITMENT AND SELECTION

You can begin a sort of "pre-training" for new employees by making sure that you incorporate key principles of your institution into job opening announcements. If yours is a facility with expectations for inquiry-based learning, make sure that potential applicants know that and are comfortable with that style. Include those requirements in the position announcement.

A lot of retraining can be avoided through careful initial personnel selection. Cranbrook's program staff has developed a very effective but non-threatening method of assessing whether a potential applicant grasps the museum's learning style. As part of a group interview with other education department staff, the candidate is asked how he/she might use some sample objects or an exhibit to help others learn. If the person is stymied, we have a clue that it might be a long training process to move that person into the museum's learning model; another person might immediately engage himself/herself with the exhibit, begin asking questions about it, and show both his/her content knowledge and teaching style.

RESPONSIBILITY FOR TRAINING

Training needs to be someone's direct responsibility in order for it to be consistently implemented. A logical place for such responsibility is with a program or education staff person because of their required visitor orientation.

PERSONNEL TO BE TRAINED

Everyone who deals with the public needs to have some training. Telephone answerers, admissions staff, public relations and development personnel, administrators, custodians, and volunteers all need to have a complete understanding of the museum. The best way to help these staff and volunteers understand exhibits and programs is to have them experience the museum as though they are visitors. Get them away from their desks or their normal duties to use the exhibits, to ask questions about them, to explain them to each other, and then to be able to pass their enthusiasm and their understanding along to others.

TRAINING FOR EXPLAINERS AND OTHER PROGRAM PERSONNEL

The most in-depth training needs to be directed toward program staff. Once a person has been hired, give him/her time to get used to the feel of the museum. Time spent just observing visitors in the exhibits is time well spent. Just fiddle with the exhibits, perhaps for a day or two or at random hours, to absorb the feel of the museum and the sense of the visitors. New employees should also observe everyone else doing programs or explaining in the exhibits, so that they can sense the variables that are possible within one learning model.

Be sure that there are content outlines for demonstrations and programs, not fully scripted but complete with goals, objectives, and essential highlights. Produce a teaching guide that focuses on the "how to's" and not just the "what's." Team a new employee with a seasoned one for the first few public efforts, and have videotapes of experienced employees working on the exhibit floor for leisurely review.

Require that staff regularly do abbreviated evaluation sessions with participants in their programs. For example, at the end of a demonstration or program, as a group or visitor is leaving, get quick comments from two or three random participants.

What one thing will the visitor remember most about this program or demonstration?

Set aside "talking time" with staff from other departments. Eat lunch together, go to a conference or meeting together, or have a beer after work.

TRAINING OF VOLUNTEERS

Don't forget volunteers in your training planning. In fact, they may require the most organized, formal training program. Because they tend to come and go more often than staff, be prepared to develop a buddy system so that you can have "old" volunteers help with training of new volunteers. Here, a training manual is a must. Also, consider videotaping good explainers at work, then have an easily accessible library of videos that new volunteers can check out for quick review.

NEEDS OF ADOLESCENTS

Teens and young teens in the museum add a dimension of excitement, and you will find that they make great explainers because they are usually so uninhibited! They do need clear direction, or you may find that the whole museum is their playground. They also like to "clump" together—and you can use that to your advantage as well. Have them work in teams, on a defined project, with definite deadlines. It is not a bad idea to have them sign a mini-contract, in which they agree to do a specific task within a given time. They need an element of fun, too, so be sure that is included.

SUBTLE TRAINING EFFORTS

One of Cranbrook's most effective ways of training staff and volunteers is during a weekly staff lunch. We don't call it a training session, but every Thursday at 12:00 noon we all go to the "penthouse" for a bring-your-own-lunch (with dessert supplied by participants on a rotating basis). This has been going on for years and is a memorable Cranbrook tradition. At 12:30, the director leads a discussion, which may be as straightforward as announcements of upcoming events, so that everyone knows what is happening, or as complex as discussion about renovation and expansion planning. This puts everyone on an even level, with common information on which to base actions. A good amount of training is already done, and people hardly know it happened! Both staff and volunteers are better informed and able to move the vision forward, and they feel part of a whole, vital organization.

REACHING UNDERREPRESENTED AUDIENCES

Not all groups feel equally comfortable in a museum setting. Properly trained staff and volunteers can ensure that non-traditional audiences feel welcome. A warm smile at the door and helpful assistance is a good beginning for everyone, but it is even more important for first-time visitors from under-represented groups. Make sure they have a floor plan or other printed material so that they get off to a good start and feel confident about finding their way around. The most success-ful way to build an audience of underrepresented groups is by word-of-mouth, but you have to help see that every visitor has a good museum experience.

DON'T MAKE TRAINING TRIVIAL

With deadlines always pressing and new activities just around the corner, it's easy to forget about training. Don't do that! Effective training is a time-saver in the long run, and it helps assure that all of a museum's employees and volunteers are visitor-centered, and that they convey the sense of the museum.

But make training fun, too. If you feel the need for a formal training session, do it late in the day—maybe over pizza. Make it something people look forward to. A new employee who is nurtured from the start through effective training will be more confident about his/her role and do a far better job for the museum.

Janet Johnson began her career as a high school teacher in Lincoln Park, Michigan, after receiving a B.S. in English from the University of Minnesota. She joined the staff at Cranbrook Institute of Science in 1978 as coordinator of community relations, and is now director of education. Located just north of Detroit in Bloomfield Hills, Michigan, the Institute is situated on a 315-acre campus of the Cranbrook Educational Community.

SETTING UP
THE BUSINESS

Who runs a science center, and what does it cost? Where does the money come from? What do you need to get started? This chapter describes science centers as businesses—mission-oriented though they are—and discusses their operations quantitatively. Most of the published material on science center structure and financing on which my comments are based comes from North American institutions. But conversations with colleagues on other continents lead me to believe that many of these considerations apply as well outside the U.S. and Canada.

Transitions pervade a science center's early years.

The universe of science centers contains many small institutions and fewer large ones. Physical size, budget, and attendance roughly parallel each other. The smallest institutions occupy several thousand square feet, serve tens of thousands of visitors, and operate on perhaps $100,000. The largest centers occupy hundreds of thousands of square feet, serve over a million people, and operate on perhaps $10 million. When I refer to "smaller" or "larger" science centers, I mean the combination of physical plant, financial resources, and people reached. On this composite scale, the newest science centers tend to be smaller (under 50,000 square feet, $2 million in operating budget, and 200,000 in annual visitors), although every so often a large, new center opens that proves the rule.

Although small and large science centers share aspects of their mission and techniques, their internal structures differ markedly, as might be expected. This makes it difficult to generalize about organizations as different as your local library is from the Library of Congress. Institutions also face a wide variety of political and economic constraints. Nonetheless, I want to point out some basic business concerns that pertain to most start-up science centers.

MANAGEMENT DYNAMICS

As not-for-profit institutions, most science centers are governed by a volunteer board of trustees to whom a professional executive director reports. The director is responsible for the performance of all other staff, paid or volunteer. In theory, the board establishes the organization's mission and regularly reviews its accomplishments and failings. The board bears legal responsibility for the conduct of the organization, including prudent fiscal management—getting enough money and spending it wisely. In theory, the director and his or her staff develop policies to meet the institution's objectives and implement those approved by the board. The director also supports and participates in the board's work, as deemed mutually desirable.[1]

In practice, lines of responsibility are not always so neatly drawn. Staff often initiate funding requests and organize the board's fundraising, and board members often initiate and develop programs. Especially in the pre-opening years of smaller centers, before there's money for staff, board members sometimes do everything: They write exhibit proposals, stuff envelopes, take demonstrations to classrooms, attend countless civic meetings, and do whatever else needs to be done, as well as performing their oversight role. Later, as the institution grows, such a "working board" may find it no longer meets the institution's needs—primarily for money. Somewhere along the line, especially when a new building is contemplated, a center faces the problem of replacing a "working board" with a "funding board." The people who have given so much time to the day-to-day operation of a fledgling center are often not the same people who can raise large amounts of capital for a

building or exhibition. The board has to change, however reluctantly. Managing this transition requires careful planning by board and staff leaders.

Transitions pervade a science center's early years. The 1986 ASTC survey showed that four out of five science centers and museums were involved in planning a major expansion. Expanding a building involves more than raising capital; you need to prepare for a hefty impact on operations. Increasing the number of visitors that can be served means adding staff to serve those visitors, increasing the revenue stream to support the staff, and perhaps changing the composition of the board to reflect altered constituencies and greater responsibilities. So expanding is akin to opening a new building. Responsibilities shift within the organization and an increasing number of "specialists" are needed to replace the "generalists" who used to carry the ball, both on the board and on the staff, perhaps even at the top.

If current trends continue, new science centers can look forward to continual growth.

If current trends in the field continue, new science centers can look forward to continual growth requiring continual structural readjustment and entrepreneurial acumen. They can also look forward to playing increasingly important roles in their community—even to contributing to its economic development,[2] which for some may prove to be an organizational theme.

Most often, science center staff are organized into three streams: operations, development (fundraising), and program, the last including separate exhibits and education functions. Sample organizational charts, size statistics, and operating budgets for institutions of various sizes from the ASTC survey report are reproduced here in Appendix B. ASTC publishes periodic surveys of typical salaries, as well.[3] There are no preferred organizational schemes, since much depends on local talent and institutional priorities.

Unlike traditional museums, science centers generally do not employ curators in a given subject matter. They rarely maintain taxonomic or historical collections and so do not employ conservators. Instead, they allocate scarce salary funds to skilled communicators, who engage scientists to assist them in interpretive projects, often on a project-by-project basis. Of course, the more staff and board members know and care about science and technology, the better they will shape their center's communication.

Where do you find entrepreneurial managers and skilled science communicators to operate your new center? In 1988 and 1989, ASTC ran a training program for people founding science centers that was attended by representatives of 38 developing institutions, 27 in the U.S. and one in each of 11 other countries. The founders, mostly staff or "working board," were a diverse group with a wide range of professional backgrounds—retired science teachers, museum administrators, senior scientists, and community activists (in the U.S., particularly members of the Junior League). These people were dedicated, energetic, willing to learn "on the job" and to ask for help from established science centers when needed. Founders, it seems, are not in short supply. And management training for nonprofit organizations is available.

But only a few of these people had the skills or desire to go into the shop to build exhibits, the hallmark of science centers. Most expected to hire knowledgeable, talented people to handle this job for them. But such people are rare, indeed. Only those who have practiced the art at existing science centers know what is involved. Trade schools and graduate schools of design don't produce them (with the notable exception of a program at the Gwent College of Higher Education in Great Britain). High school science teachers who tinker a lot come closer in skills and experience to the kind of person needed to develop and maintain a lively exhibit floor. New science centers often have to "grow their own" exhibit builders, and the time involved may slow the institution's development. The other kinds of skills required to start and run a center—administrative, financial, teaching, even science—are easier to find.

FUNDING PATTERNS

For the wide range of institutions ASTC surveyed in 1986, the pattern of operating support generally thought to apply to museums held true. Most institutions received about a third of their annual funding from a government source (local, state, or federal). Another third was provided by individuals and corporations in the form of grants and charitable contributions. And the remaining third was earned—as admissions, fees for programs or other services, and to a lesser degree, investment and store income. (Most often memberships are also counted as earned income.)

For new science centers, the general pattern of operating funding reverses. Among the surveyed centers that opened after 1979, about two-thirds of income was earned. Many centers that have opened after the 1986 survey claim still higher percentages of earned income. And some developing centers say they intend to survive exclusively on admissions and fees. Their strategy is to raise one-time capital funds to open and equip their building, and then to live off the gate.

It has not been established that science centers can thrive on earned income alone, although the trend, even among older centers, is in the direction of increased reliance on fees for a wide variety of services rather than governmental or philanthropic support. To date, the largely middle class audience at science centers has been willing to pay modest admission, program, and sometimes parking fees. It seems safe to assume that they will continue to do so, although we don't know how far their willingness can be stretched. But counting on the gate presents problems.

First, there's finding money for growth. If you want to explore an unusual subject area or teaching technique, or to invest in a highly labor-intensive program like teacher education, you will need to invest in people and equipment that won't produce admissions receipts very quickly, if at all. Unable to justify the expense of your experiment, you may be forced to abandon it. Second, relying on the market may make reaching beyond the accustomed audience difficult. It takes more time and money to market to people who may not be accustomed to visiting science centers. You may not be able to charge your core audience enough to pay for programs for others. Some other kind of subsidy may become necessary. In any event, you are still likely to have to fund capital projects that need large cash outlays from sources other than a steady accumulation at the gate.

PLANNING AND BUDGETING

As soon as the leadership group has developed an informed, consistent sense of your center's mission, you should write out your program—your description for what happens in the building for all your various audiences—and develop a corresponding operating budget, projected over three to five years. Preparing such an "operating plan" helps you clarify your options; having a well-developed financial statement helps supporters gain confidence in your project. In the gestational years, you may need to revise the plan periodically; even the revisions teach you something about what you need to do to achieve your goals. While there are no standard formulas for program and budget statements, general accounting principles apply.[4]

For purposes of planning, and bearing in mind the great diversity of science centers, you may want to compare your projected operating expenses to others'. In 1986, most science centers expended about $9 per visitor to keep open and running. (Earned income amounted to $5 per visitor for new science centers, $3 for the field in general.) Often, very small and very large institutions spent less; the former do away with luxuries, like two-color brochures, while the latter enjoy some economies of scale. Both very small and very large centers devote a slightly higher proportion of their facility to exhibits, which serve a larger number of visitors than other kinds of programming, and this affects their cost-per-visitor ratio.

Before using this ratio to measure your "efficiency," note that science centers vary in the way they account for exhibit development costs. Some centers include ongoing exhibit development in operations and so have a higher cost per visitor. Other centers use capital funds for occasional exhibit building and so have a lower ratio. Also, in the New York metropolitan area, it costs less for a child to spend an hour at a science center than at a public school. This helps justify a well-reasoned collaboration with the school community.

You will also need to develop a separate capital budget to raise money for the cost of renovations or new construction to house your center. Much of the figuring can be done by your architect, in response to the operating plan you have developed. But there are many costs associated with a capital project that go beyond the architectural, and if you don't include them in your request for capital funding, you will find it much harder to pay for them later. In addition to the cost of land, site preparation, construction, and architectural and engineering fees, a capital budget should include:

- fundraising costs

- additional legal and office costs

- interior design, building signage, and cabinetwork for lobby, store, classrooms, library

- office furniture and equipment

- woodworking and machine shop and building maintenance equipment

- construction management contract, if desirable

- salaries for pre-opening staff

Many centers also include money for exhibits in the start up capital budget, and they sometimes ask the sponsors of various exhibitions or galleries to provide an endowment to support ongoing interpretation of "their" exhibits. (Of course, sponsorship should not extend as far as editorial control.) Interpretation can mean special equipment maintenance, demonstration supplies, or "explainer" salaries, as the case may be. If you can raise endowment money in this way, the capital campaign can help ease your annual operational burden.[5]

In writing both operating and capital budgets, you will make assumptions about how much revenue can be obtained from various sources. When the financial stakes are high, most developing science centers ask fundraising specialists to check these assumptions—to interview potential supporters and analyze market constraints in a "funding feasibility study"—

before committing themselves to a specific course of action. Although there is copious literature about fundraising in general, much depends on local circumstances, and you can learn a lot by talking with people who know the local scene. Don't simply guess how much money is out there; guesstimates made by well-meaning but inexperienced trustees have left quite a few developing institutions embarrassingly short of the mark.

Operating Budget Facts and Figures. For those people involved in writing budgets, here are some rules of thumb. Remember, the great diversity among science centers means that you should apply these with a grain of salt:

- Science centers typically budget a little over half their operating funds for staff salaries and benefits, not including volunteers. (When requesting financial support, you can sometimes count the value of volunteer labor as in-kind matching.)

- Most centers charge admission (typically less than the price of a movie) but make some provision for free admission for special audiences and, of course, members. Total attendance in 1986 averaged 10 people per square foot of exhibit space, as described in an earlier chapter.

- Stores or gift shops do not usually make much money when all costs, including salaries, are subtracted from revenues. Auditors advise against running a store with volunteers, but if you do, profit goes up.

- Few centers run their own food service, finding it more economical to engage a contractor. Except for large centers, food service is usually not a significant source of income.

- Contributors and visitors alike want to think that the

money they give or pay goes into science programs and exhibits, not fundraising events and overhead. Be prepared to track fundraising costs and fundraising income annually; you will be questioned about them. The accepted wisdom in the museum field is that these costs should not exceed 25 percent of contributions in a normal year.[6]

MAKING IT ALL WORK

Say that you are lucky. Your fundraising counsel ascertains that all the money you need is out there, waiting to be gathered. You will likely spend a minimum of five years to complete the capital campaign, build the building, create exhibits, and finally open the doors.

More often than not, it will cost more money and take more time to complete construction than anticipated, and you will be tempted to raid the exhibit budget to pay for bricks and mortar. Opening incomplete may damage public relations. You may have to increase the level of support you devote to on-going exhibit development.

In some places, people mount demonstration projects before planning to build anything at all. The project—usually a temporary or traveling exhibition of interactive displays—serves to test the notion of a science center for the locale. Based on the public and political response to the project, planners proceed to develop physical and financial descriptions of a science center to come. And they may succeed in attracting some "pilot" financial support.

Many science centers have elected a different, low-overhead course of development. Starting with a storefront or a rent-free municipal building, a small group of dedicated organizers has quickly begun operations on a modest scale, meeting the

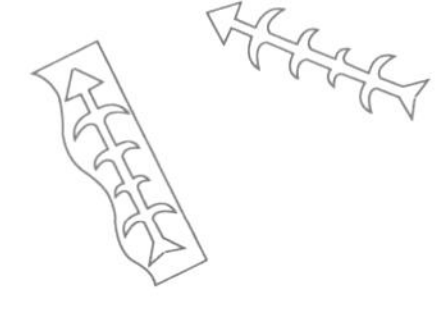

public on weekends and practicing teaching techniques with selected school groups. These centers have chosen to "grow as you go," expecting support for the concept of the center to increase over time. They have gradually built a repertoire of interpretive techniques and a local following, and then considered the possibility of doing more.

Whatever you start with—a storefront, half a renovated hall, or a new building—the key to a successful start-up phase is reaching a critical mass of programming: enough to attract and satisfy a sufficient number of visitors to pay for the staff required to serve them, and for whatever fundraising effort is needed to make up revenue beyond admissions. Since it is hard to start a fundraising effort from scratch, you will probably need to look for a core of support for the first three to five years, a dependable subsidy that will allow you to develop your own financial muscle.

Some newly opened science centers report large attendance in the first year and then a decline after the novelty wears off. Others report steady or increasing attendance in the first years. ASTC is collecting information on attendance at newly opened science centers; there may be patterns that can guide your planning. In the meantime, it would be prudent to project attendance conservatively, and then hope for a surplus!

Since admissions revenues mean a lot, new science centers pay attention to marketing. Don Adams, an authority on museum public relations,[7] says that most museum marketers have little formal background and borrow strategies from other organizations, perhaps missing the mark for their own institutions. Marketing doesn't mean heavy advertising and promotion, but rather ensuring that visitors' comments prompt more new and repeat visitors. You need to find out what your visitors think of your place and check whether their expectations, set by word-of-mouth, are being met. Your educational mission should drive your marketing message; you shouldn't make promises to visitors that your center cannot keep. But message and program may both need to be adjusted, using feedback collected systematically from the public, to maximize the power of word of mouth to bring visitors to your place.

Many science centers have elected to "grow as you go," expecting support for the concept of the center to increase over time.

Most people founding science centers spend years of effort in anticipation of opening day. And then the doors open, visitors come, and their jobs are transformed. Opening is just a beginning; serving the day-to-day needs of the visiting public and strengthening a young institution bring new challenges. Your work of the last few years has prepared you to meet them, though, and you are buoyed up by your visitors' interest and pleasure and your own well-earned sense of accomplishment.

Notes

1. For a concise guide to board functioning, see Joseph Weber, *Managing the Board of Directors*. The standard work is Alan Ullberg and Patricia Ullberg, *Museum Trusteeship*. See also Stephen E. Weil, "A Checklist of Legal Considerations for Museums."

2. For one discussion of economic impact, see Sandra Wilcoxon, "Measuring Your Impact."

3. Association of Science-Technology Centers, *1990 ASTC Salary Survey*.

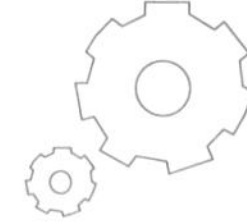

4. See Kent Chabotar, "Cost Analysis in Schools and Other Nonprofits: A Management Perspective."

5. For an introduction to capital campaigning, see David Nichols, "Managing the Transition into the Campaign," and Margaret Maxwell, "You Can Build a Campaign." A classic article on development is Carl Shaver, "The Rights and Rituals of Fund Raising."

6. See Peter Ames, "Measures of Merit?"

7. See G. Donald Adams, *Museum Public Relations*.

VIEWPOINTS

In an article about what businesses can learn from nonprofits, management expert Peter Drucker discusses how successful nonprofits use their mission to organize the performance of board, staff, and volunteers. Don Adams and John Boatright describe the background against which institutional communications are received by the public, and marketing techniques used to dramatically improve attendance at the Henry Ford Museum and Greenfield Village complex.

WHAT BUSINESS CAN LEARN FROM NONPROFITS *Peter F. Drucker*

The Girl Scouts, the Red Cross, the pastoral churches—our nonprofit organizations—are becoming America's management leaders. In two areas, strategy and the effectiveness of the board, they are practicing what most American businesses only preach. And in the most crucial area—the motivation and productivity of knowledge workers—they are truly pioneers, working out the policies and practices that business will have to learn tomorrow.

Few people are aware that the nonprofit sector is by far America's largest employer. Every other adult—a total of 80 million plus people—works as a volunteer, giving on average nearly five hours each week to one or several nonprofit organizations. This is equal to 10 million full-time jobs. Were volunteers paid, their wages, even at minimum rate, would amount to some $150 billion, or 5 percent of GNP. And volunteer work is changing fast. To be sure, what many do requires little skill or judgment: collecting in the neighborhood for the Community Chest one Saturday afternoon a year, chaperoning youngsters selling Girls Scout cookies door to door, driving old people to the doctor. But more and more volunteers are becoming "unpaid staff," taking over the professional and managerial tasks in their organizations.

Not all nonprofits have been doing well, of course. A good many community hospitals are in dire straits. Traditional churches and synagogues of all persuasions—liberal, conservative, evangelical, fundamentalist—are still steadily losing members. Indeed, the sector overall has not expanded in the last 10 or 15 years, either in terms of the money it raises (when adjusted for inflation) or in the number of volunteers. Yet in its productivity, in the scope of its work and in its contribution to American society, the nonprofit sector has grown tremendously in the last two decades.

The Salvation Army is an example. People convicted to their first prison term in Florida, mostly very poor black or Hispanic youths, are now paroled into the Salvation Army's custody—about 25,000 each year. Statistics show that if these young men and women go to jail the majority will become habitual criminals. But the Salvation Army has been able to rehabilitate 80 percent of them through a strict work program run largely

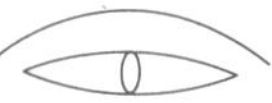

by volunteers. And the program costs a fraction of what it would to keep the offenders behind bars.

Underlying this program and many other effective nonprofit endeavors is a commitment to management. Twenty years ago, management was a dirty word for those involved in nonprofit organizations. It meant business, and nonprofits prided themselves on being free of the taint of commercialism and above such sordid considerations as the bottom line. Now most of them have learned that nonprofits need management even more than business does, precisely because they lack the discipline of the bottom line. The nonprofits are, of course, still dedicated to "doing good." But they also realize that good intentions are no substitute for organization and leadership, for accountability, performance, and results. Those require management and that, in turn, begins with the organization's mission.

> **Good intentions are no substitute for organization and leadership, for accountability, performance, and results.**

As a rule, nonprofits are more money-conscious than business enterprises are. They talk and worry about money much of the time because it is so hard to raise and because they always have so much less of it than they need. But nonprofits do not base their strategy on money, nor do they make it the center of their plans, as so many corporate executives do. "The businesses I work with start their planning with financial returns," says one well-known CEO who sits on both business and nonprofit boards. "The nonprofits start with the performance of their mission."

Starting with the mission and its requirements may be the first lesson business can learn from successful nonprofits. It focuses the organization on action. It defines the specific strategies needed to attain the crucial goals. It creates a disciplined organization. It alone can prevent the most common degenerative disease of organizations, especially large ones: splintering their always limited resources on things that are "interesting" or look "profitable" rather than concentrating them on a very small number of productive efforts.

The best nonprofits devote a great deal of thought to defining their organization's mission. They avoid sweeping statements full of good intentions and focus, instead, on objectives that have clear-cut implications for the work their members perform—staff and volunteers both. The Salvation Army's goal, for example, is to turn society's rejects—alcoholics, criminals, derelicts—into citizens. The Girl Scouts help youngsters become confident, capable young women who respect themselves and other people. The Nature Conservancy preserves the diversity of nature's fauna and flora. Nonprofits also start with the environment, the community, the "customers" to be; they do not, as American businesses tend to do, start with the inside, that is, with the organization or with financial returns.

Willowcreek Community Church in South Barrington, Illinois, outside Chicago, has become the nation's largest church—some 13,000 parishioners. Yet it is barely 15 years old. Bill Hybels, in his early twenties when he founded the church, chose the community because it had relatively few churchgoers, though the population was growing fast and churches were plentiful. He went from door to door asking, "Why don't you go to church?" Then he designed a church to answer the potential customers' needs: for instance, it offers full services on Wednesday evening because many working parents need Sunday to spend with their children. Moreover, Hybels

continues to listen and react. The pastor's sermon is taped while it is being delivered and instantly reproduced so that parishioners can pick up a cassette when they leave the building because he was told again and again, "I need to listen when I drive home or drive to work so that I can build the message into my life." But he was also told: "The sermon always tells me to change my life but never how to do it." So now every one of Hybel's sermons ends with specific action recommendations.

A well-defined mission serves as a constant reminder of the need to look outside the organization not only for "customers" but also for measures of success. The temptation to content oneself with the "goodness of our cause"—and thus to substitute good intentions for results—always exists in nonprofit organizations. It is precisely because of this that the successful and performing nonprofits have learned to define clearly what changes outside the organization constitute "results" and to focus on them.

The experience of one large Catholic hospital chain in the Southwest shows how productive a clear sense of mission and a focus on results can be. Despite the sharp cuts in medicare payments and hospital stays during the past eight years, this chain has increased revenues by 15 percent (thereby managing to break even) while greatly expanding its services and raising both patient-care and medical standards. It has done so because the nun who is its CEO understood that she and her staff are in the business of delivering health care (especially to the poor), not running hospitals.

As a result, when health care delivery began moving out of hospitals for medical rather then economic reasons about ten years ago, the chain promoted the trend instead of fighting it. It founded ambulatory surgery centers, rehabilitation centers,

X-ray and lab networks, HMOs, and so on. The chain's motto was: "If it's in the patient's interest, we have to promote it; it's then our job to make it pay." Paradoxically, the policy has filled the chain's hospitals; the free-standing facilities are so popular they generate a steady stream of referrals.

A well-defined mission serves as a constant reminder of the need to look outside the organization not only for "customers" but also for measures of success.

This is, of course, not so different from the marketing strategy of successful Japanese companies. But it is very different indeed from the way most Western businesses think and operate. And the difference is that the Catholic nuns—and the Japanese—start with the mission rather than with their own rewards, and with what they have to make happen outside themselves, in the marketplace, to deserve a reward.

Finally, a clearly defined mission will foster innovative ideas and help others understand why they need to be implemented—however much they fly in the face of tradition. To illustrate, consider the Daisy Scouts, a program for five-year-olds which the Girls Scouts initiated a few years back. For 75 years, first grade had been the minimum age for entry into a Brownie troop, and many Girl Scout councils wanted to keep it that way. Others, however, looked at demographics and saw the growing numbers of working women with "latch key" kids. They also looked at the children and realized that they were far more sophisticated than their predecessors a generation ago (largely thanks to TV).

Today the Daisy Scouts are 100,000 strong and growing fast. It is by far the most successful of the many programs for preschoolers that have been started these last 20 years, and far more successful than any of the very expensive government programs. Moreover, it is so far the only program that has seen these critical demographic changes and children's exposure to long hours of TV viewing as an opportunity.

Many nonprofits now have what is still the exception in business—a functioning board. They also have something even rarer: a CEO who is clearly accountable to the board and whose performance is reviewed annually by a board committee. And they have what is rarer still: a board whose performance is reviewed annually against preset performance objectives. Effective use of the board is thus a second area in which business can learn from the nonprofit sector.

In U.S. law, the board of directors is still considered the "managing" organ of the corporation. Management authors and scholars agree that strong boards are essential and have been writing to that effect for more than 20 years, beginning with Myles Mace's pioneering work.[1] Nevertheless, the top managements of our large companies have been whittling away at the directors' role, power, and independence for more than half a century. In every single business failure of a large company in the last few decades, the board was the last to realize that things were going wrong. To find a truly effective board, you are much better advised to look in the nonprofit sector than in our public corporations.

In part, this difference is a product of history. Traditionally, the board has run the shop in nonprofit organizations—or tried to. In fact, it is only because nonprofits have grown too big and complex to be run by part-time outsiders, meeting for three hours a month, that so many have shifted to professional management. The American Red Cross is probably the largest nongovernmental agency in the world and certainly one of the most complex. It is responsible for worldwide disaster relief; it runs thousands of blood banks as well as the bone and skin banks in hospitals; it conducts training in cardiac and respiratory rescue nationwide; and it gives first-aid courses in thousands of schools. Yet it did not have a paid chief executive until 1950, and its first professional CEO came only with the Reagan era.

Many nonprofits have what is still the exception in business—a functioning board.

But however common professional management becomes—and professional CEOs are now found in most nonprofits and all the bigger ones—nonprofit boards cannot, as a rule, be rendered impotent the way so many business boards have been. No matter how much nonprofit CEO's would welcome it—and quite a few surely would—nonprofit boards cannot become their rubber stamp. Money is one reason. Few directors in publicly held corporations are substantial shareholders, whereas directors on nonprofit boards very often contribute large sums themselves, and are expected to bring in donors as well. But also, nonprofit directors tend to have a personal commitment to the organization's cause. Few people sit on a church vestry or on a school board unless they deeply care about religion or education. Moreover, nonprofit board members typically have served as volunteers themselves for a good many years and are deeply knowledgeable about the organization, unlike outside directors in a business.

Precisely because the nonprofit board is so committed and active, its relationship with the CEO tends to be highly contentious and full of potential for friction. Nonprofit CEOs complain that their board "meddles." The directors, in turn, complain that management "usurps" the board's function.

This has forced an increasing number of nonprofits to realize that neither board nor CEO is "the boss." They are colleagues, working for the same goal but each having a different task. And they have learned that it is the CEO's responsibility to define the tasks of each, the board's and his or her own.

For example, a large electric co-op in the Pacific Northwest created ten board committees, one for every member. Each has a specific work assignment: community relations, electricity rates, personnel, service standards, and so on. Together with the co-op's volunteer chairman and its paid CEO, each of these one-person committees defines its one-year and three-year objectives and the work needed to attain them, which usually requires five to eight days a year from the board member. The chairman reviews each member's work and performance every year, and a member whose performance is found wanting two years in a row cannot stand for reelection. In addition, the chairman, together with three other board members, annually reviews the performance of the entire board and of the CEO.

The key to making a board effective is not to talk about its function but to organize its work.

The key to making a board effective, as this example suggests, is not to talk about its function but to organize its work. More and more nonprofits are doing just that, among them half a dozen fair-sized liberal arts colleges, a leading theological seminary, and some large research hospitals and museums. Ironically, these approaches reinvent the way the first nonprofit board in America was set up 300 years ago: the Harvard University Board of Overseers. Each member is assigned as a "visitor" to one area in the university—the Medical School, the Astronomy Department, the investment of the endowment—and acts both as a source of knowledge to that area and as a critic of its performance. It is a common saying in American academia that Harvard has the only board that makes a difference.

The weakening of the large corporation's board would, many of us predicted (beginning with Myles Mace), weaken management rather than strengthen it. It would diffuse management's accountability for performance and results; and indeed, it is the rare big-company board that reviews the CEO's performance against preset business objectives. Weakening the board would also, we predicted, deprive top management of effective and credible support if it were attacked. These predictions have been borne out amply in the recent rash of hostile takeovers.

To restore management's ability to manage we will have to make boards effective again—and that should be considered a responsibility of the CEO. A few first steps have been taken. The audit committee in most companies now has a real rather than a make-believe job responsibility. A few companies—though so far almost no large ones—have a small board committee on succession and executive development, which regularly meets with senior executives to discuss their performance and their plans. But I know of no company so far where there are work plans for the board and any kind of review of the board's performance. And few do what the larger nonprofits now do routinely: put a new board member through systematic training.

Nonprofits used to say, "We don't pay volunteers so we cannot make demands upon them." Now they are more likely to say, "Volunteers must get far greater satisfaction from their accomplishments and make a greater contribution precisely because they do not get a paycheck." The steady transforma-

tion of the volunteer from well-meaning amateur to trained, professional, unpaid staff member is the most significant development in the nonprofit sector—as well as the one with the most far-reaching implications for tomorrow's businesses.

A Midwestern Catholic Diocese may have come furthest in this process. It now has fewer than half the priests and nuns it had only 15 years ago. Yet it has greatly expanded its activities—in some cases, such as help for the homeless and for drug abusers, more than doubling them. It still has many traditional volunteers like the Altar Guild members who arrange flowers. But now it is also being served by some 2,000 part-time unpaid staff who run the Catholic charities, perform administrative jobs in parochial schools, and organize youth activities, college Newman Clubs, and even some retreats.

A similar change has taken place at the First Baptist Church in Richmond, Virginia, one of the largest and oldest churches in the Southern Baptist Convention. When Dr. Peter James Flamming took over five years ago, the church had been going downhill for many years, as is typical of old, inner-city churches. Today it again has 4,000 communicants and runs a dozen community outreach programs as well as a full complement of in-church ministries. The church has only nine paid full-time employees. But of its 4,000 communicants, 1,000 serve as unpaid staff.

This development is by no means confined to religious organizations. The American Heart Association has chapters in every city of any size throughout the country. Yet its paid staff is limited to those at national headquarters, with just a few traveling troubleshooters serving the field. Volunteers manage and staff the chapters, with full responsibility for community health education as well as fundraising.

These changes are, in part, a response to need. With close to half the adult population already serving as volunteers, their overall number is unlikely to grow. And with money always in short supply, the nonprofits cannot add paid staff. If they want to add to their activities—and needs are growing—they have to make volunteers more productive, have to give them more work and more responsibility. But the major impetus for the change in the volunteer's role has come from the volunteers themselves.

More and more volunteers are educated people in managerial or professional jobs—some preretirement men and women in their fifties, even more babyboomers who are reaching their mid-thirties or forties. These people are not satisfied with being helpers. They are knowledge workers in the jobs in which they earn their living, and they want to be knowledge workers in the jobs in which they contribute to society—that is, their volunteer work. If nonprofit organizations want to attract and hold them, they have to put their competence and knowledge to work. They have to offer meaningful achievement.

Many nonprofits systematically recruit for such people. Seasoned volunteers are assigned to scan the newcomers—the new member in a church or synagogue, the neighbor who collects for the Red Cross—to find those with leadership talent and persuade them to try themselves in more demanding assignments. Then senior staff (either a full-timer on the payroll or a seasoned volunteer) interviews the newcomers to assess their strengths and place them accordingly. Volunteers may also be assigned both a mentor and a supervisor with whom they work out their performance goals. These advisers are two different people, as a rule, and both, ordinarily, volunteers themselves.

The Girl Scouts, which employs 730,000 volunteers and only 6,000 paid staff for 3½ million girl members, works this way. A volunteer typically starts by driving youngsters once a week to a meeting. Then a more seasoned volunteer draws her into other work—accompanying Girl Scouts selling cookies door-to-door, assisting a Brownie leader on a camping trip. Out of this step-by-step process evolve the volunteer boards of the local councils and, eventually, the Girl Scouts governing organ, the National Board. Each step, even the very first, has its own compulsory training program, usually conducted by a woman who is herself a volunteer. Each has specific performance standards and performance goals.

What do these unpaid staff people themselves demand? What makes them stay—and, of course, they can leave at any time. Their first and most important demand is that the nonprofit have a clear mission, one that drives everything the organization does. A senior vice president in a large regional bank has two small children. Yet she just took over as chair of the state chapter of Nature Conservancy, which finds, buys, and manages endangered natural ecologies. "I love my job," she said, when I asked her why she took on such heavy additional work, "and of course the bank has a creed. But it doesn't really know what it contributes. At Nature Conservancy, I know what I am here for."

The transformation of the volunteer from well-meaning amateur to trained, professional, un-paid staff member is the most significant development in the nonprofit sector.

The second thing this new breed requires, indeed demands, is training, training, and more training. And, in turn, the most effective way to motivate and hold veterans is to recognize their expertise and use them to train newcomers. Then these knowledge workers demand responsibility—above all, for thinking through and setting their own performance goals. They expect to be consulted and to participate in making decisions that affect their work and the work of the organization as a whole. And they expect opportunities for advancement, that is, a chance to take on more demanding assignments and more responsibility as their performance warrants. That is why a good many nonprofits have developed career ladders for their volunteers.

Supporting all this activity is accountability. Many of today's knowledge-worker volunteers insist on having their performance reviewed against preset objectives at least once a year. And increasingly, they expect their organizations to remove nonperformers by moving them to other assignments that better fit their capacities or by counseling them to leave. "It's worse than the Marine Corps boot camp," says the priest in charge of volunteers in the Midwestern diocese, "but we have 400 people on the waiting list." One large and growing Midwestern art museum requires of its volunteers—board members, fundraisers, docents, and the people who edit the museum's newsletter—that they set their goals each year, appraise themselves against these goals each year, and resign when they fail to meet their goals two years in a row. So does a fair-sized Jewish organization working on college campuses.

These volunteer professionals are still a minority, but a significant one—perhaps a tenth of the total volunteer population. And they are growing in numbers and, more important, in their impact on the nonprofit sector. Increasingly, nonprofits say what the minister in a large pastoral church says: "There is no laity in this church; there are only pastors, a few paid, most unpaid."

This move from nonprofit volunteer to unpaid professional may be the most important development in American society today. We hear a great deal about the decay and dissolution of family and community and about the loss of values. And, of course, there is reason for concern. But the nonprofits are generating a powerful countercurrent. They are forging new bonds of community, a new commitment to active citizenship, to social responsibility, to values. And surely what the non-profit contributes to the volunteer is as important as what the volunteer contributes to the nonprofit. Indeed, it may be fully as important as the service, whether religious, educational, or welfare related, that the nonprofit provides in the community.

This development also carries a clear lesson for business. Managing the knowledge worker for productivity is the challenge ahead for American management. The nonprofits are showing us how to do that. It requires a clear mission, careful placement and continuous learning and teaching, management by objectives and self-control, high demands but corresponding responsibility, and accountability for performance and results.

There is also, however, a clear warning to American business in this transformation of volunteer work. The students in the program for senior and middle-level executives in which I teach work in a wide diversity of businesses: banks and insurance companies, large retail chains, aerospace and com-puter companies, real estate developers, and many others. But most of them also serve as volunteers in nonprofits—in a church, on the board of the college they graduated from, as scout leaders, with the YMCA or the Community Chest or the local symphony orchestra. When I ask them why they do it, far too many give the same answer: Because in my job there isn't much challenge, not enough achievement, not enough responsibility; and there is no mission, there is only expediency.

Notes

1. A good example is Myles Mace, "The President and the Board of Directors," *Harvard Business Review*, March-April 1972, p. 37.

Peter Drucker is the Marie Rankin Clarke Professor of Social Sciences and Management at the Claremont Graduate School in Claremont, California, which has named its management center after him.

Reprinted with permission from the Harvard Business Review, July-August 1989, Harvard Business School Press.

THE SELLING OF THE MUSEUM 1986

G. Donald Adams and John Boatright

What if somebody told you that you were in show business? Your reaction might be somewhere between revulsion and a quizzical look. The very idea that a museum with high standards, irreplaceable collections, and professional staff should be compared to the tawdry world of show business is a thought that could be unconscionable. But the fact is, museums are competing for people's leisure time and money. And in that regard they compete in an arena that runs the gamut from VCRs and home entertainment centers to the best that Broadway, Hollywood, and Madison Avenue can deliver.

The world is changing rapidly, worldwide communication is instantaneous, and the average consumer today is significantly more sophisticated than his counterpart even 20 years ago. Consider the fact that in 1960 half of the people in the United States had not been more than 200 miles away from home. Or that in 1970 half of the people in the United States had never been on an airplane. Now you can fly from New York to Miami for $49, or coast to coast for $99. Consider that we are only 14 years away from the year 2000, when most likely there will be a level of automation, miniaturization, and communication that most of us can only relate to in Buck Rogers' terms. Those of us who are going to succeed in the marketplace must learn to function in a society that is replete with hard-nosed, bottom-line-oriented business people, who daily hone and fine-tune plans designed to eat the competition's lunch.

Ah, you say, but we are a not-for-profit organization. Really? All of you who have sufficient budgets for the operation and promotion of your organization please go on to the next article. The majority of us are finding that our endowments, membership contributions, and revenue-generating sources are simply inadequate to meet the needs   of our organizations. So we, like all other businesses, need a business plan designed to maximize return on our expenditures.

The probability is that most of you did not major in business, have had sketchy backgrounds in marketing, and know significantly more about the arts and humanities than you do about segmentation strategies. You are not alone. The fact is that many people in museums have become responsible for advertising, public relations, admissions, and merchandise sales, with no formal training in any of these disciplines.

If you find yourself in this situation, a good first step in understanding marketing is to buy a book on the subject that includes how to write a marketing plan, with particular emphasis on situation analysis. It will help you to think through the factors affecting your institution and the strategies that will help you compete successfully in the marketplace. Competition is fierce, the environment highly competitive, and your target customer is bombarded with daily messages in an effort to persuade, cajole, or entice consumers to act or react in a prescribed manner. What follows is a brief overview of the environment in which you are operating.

CONSUMER TRENDS

One of the most widely read predictors of society's behavioral themes and patterns is the _Yankelovich, Skelly and White Monitor._ What does the _Monitor_ have to say about the consumer of the 1980s? This is the time of the smart shopper.

Cost efficiency is a major concern. Quality will become more and more important. The consumer will be very pragmatic, analyzing purchases on a case-by-case basis and buying less on impulse. This will carry over into the decisions made about leisure time as options become greater and price incentives become more frequent. Value is the key operative word as we move toward the 1990s. "Me-ism," that all-encompassing trend to please oneself first and foremost, isn't going away, but it is getting weaker.

The *Monitor* forecasts that there will be a sharp return to commitment and family. This is evidenced on one hand by the leveling off of the divorce rate and on the other hand by a rise in family togetherness. There is evidence of an emerging trend toward old-fashioned ideas and nostalgia that should continue into the next decade. This is coupled with another finding showing a trend toward local identification. People increasingly demonstrate a sense of loyalty to where they live, and thus there is greater opportunity for the use of local identification, tying your product offering and its relevance to the lives of those in your region. Word-of-mouth and networking, always critical, are forecast to become even more important ways of endorsement, communication, and information. In addition to these social issues at work today, demographic trends also will shape the consumer markets of the late 1980s and '90s.

DEMOGRAPHIC TRENDS

In any marketing program it is essential that the target market be delineated in demographic terms. Just as the level of sophistication has changed dramatically in the last 20 years among consumers, so has the demography.

Two-thirds of women between the ages of 20 and 64 work outside the home, and two-thirds of women with school-age children are working or looking for work. They see the world differently than their mothers did, and the older generation is giving way to a newer one that accepts—even expects—women to work.

If your idea of the typical American family is a husband and wife with two children, with the husband working and the wife at home, think again. According to the 1980 census, only 26 percent of the U.S. households fit this mold. In fact, more than 50 percent of all households are now comprised of only one or two persons, with one in four households headed by singles. Not only does this affect disposable income, but it also has a dramatic bearing on the quality and quantity of leisure time available for an individual or family.

In our present decade, the teenage market will shrink, as will the 20-to 30-year-old age groups. At the same time, the 30-to 34-year-old segment will increase by 23 percent and the 35-to 49-year-old segment will grow by 42 percent. This certainly bodes well for those who can capture the imagination and interest of this more mature growth market.

Just as the level of sophistication has changed dramatically in the last 20 years among consumers, so has the demography.

Smart shoppers, "me-ism," return to family orientation, local identification, nostalgia, working women, and aging baby boomers are all factors that must be considered in developing a marketing plan and strategies. But consider for a moment the environment in which you will be competing and the task of arriving at the most competitive stance you can take against your target audience.

DEVELOPING A COMMUNICATIONS STRATEGY

One of the most visible, and conceivably one of the easiest, areas for you to control is the communication generated by your organization to its various publics. Remember, you are in a highly competitive environment. Your prospects are being buffeted from all sides by a variety of consumer messages. Al Ries and Jack Trout make some sobering observations in their book, *Positioning: The Battle for the Mind* (New York: McGraw-Hill, 1982). The average consumer is confronted by more than 200,000 ad messages a year—200,000! Ninety-six percent of American homes have at least one television set, and during the 50-some hours of television the average family watches each week, they see more than 600 commercials. And then, of course, there are radio spots, newspaper and magazine ads, billboards, bus signs, matchbook covers, you name it.

We live with an overdose of communication. When consumers go to the average supermarket, they can choose from more than 120,000 products and brands that are displayed. In one single product category—cigarettes—there are now more than 175 brands. How in the world do consumers cope with all of this? They don't. The human brain is a totally inadequate container for all of this information. As an example, try to recite all of the Ten Commandments, the names of the seven dwarfs, or the seven danger signals of cancer. You probably can't. And neither can your consumers. They simplify things and categorize and, therefore, see what they expect to see. Give them two paintings—one signed by Schwartz and one signed by Picasso—and listen to what they say. In their perception of the world around them, consumers screen and reject much of what is offered and accept only that which matches their prior knowledge or experience.

So how do you make sure that your communication isn't rejected? Ries and Trout say it's positioning. In the 1960s, the unique selling proposition was the key. In the '70s it was image. And in the '80s it's positioning. NASA refers to a "window in space." Positioning is finding that window, that space in the consumer's mind—that right time and right circumstance when effective communication will take place.

IBM didn't invent the computer. Sperry Rand did. But IBM was the first company to build a computer that was positioned in the minds of the public. And being first is perhaps the easiest position to establish. Who was the first person to walk on the moon? Neil Armstrong. Who was the second? Who was the first person to fly solo across the Atlantic? Charles Lindburgh. Who was the second? What's the first thing that comes to mind in photography? Kodak. In rental cars? Hertz. Cola? Harder to say these days because of changing products and positioning. Being biggest, most expensive, or least expensive are other positions.

An understanding of a museum's position in the marketplace should evolve from doing your homework. And if you've gone through the discipline of preparing a marketing plan, and specifically a situation analysis, you should have a pretty good grasp of your product, the competition, and how they both relate to your defined audiences. Following are some principles that are useful both in defining the position you want to establish with the consumer and in communicating it.

Be honest. Don't try to be something that you are not. It is amazing how many times people charged with communications want to exaggerate, bend the truth, and sometimes even create false perceptions in an effort to make their institution more attractive. Remember the rising importance of word-of-mouth advertising and how it can help you or hurt you. Your consumer will have a pre-visit expectation of your product, which will or will not match up with the impressions he or she will communicate to friends after the visit. Make sure you make a promise you can keep.

Be unique. What is it that you offer the consumer that cannot be found anywhere else? In your situation analysis, it is possible that you found you were at parity with many other institutions that offered collections and programs similar to yours. Your strategy may have been to provide an interpretation of this program that is different from the rest. The need to be communicated front and center is how you are different. If you have excellent collections, terrific. Your holdings provide a foundation upon which to build distinction in the marketplace through interpretation, exhibits, special events, or other means. Find something that is unique—and marketable—and build upon it.

If you can't be unique, be convenient. Maybe you're not the only one offering your specific product in the United States, but you may be a whole lot more accessible to the great Midwest population centers than a museum offering the same type of program on the east or west coast. Sometimes comparing yourself to a known product can help you "borrow" instant prestige from another organization.

Know your audience and speak to its interests.

Don't try to be everything to everybody. Know your audience and speak to its interests. It's going to be very difficult for you to please all the people all the time, and limited budgets almost always dictate a somewhat targeted approach toward specific audience segments. These segments could be young, old, upscale, worldly, or less urbane. Find out who

your current customers are, identify your prime prospects and speak directly to their interests. Continue to sell to your strengths. Even if you have a new accession that is exciting and warrants publicity, be sure that the stories you release are consistent with the museum's overall focus and that each release reinforces your ongoing communications effort. Be careful not to fragment your communication by sending out mixed messages—by sending out a release that positions the organization as a state museum one day and as a museum that focuses only on the history of the textile industry on another day.

Project a personality. You may appeal to the curious, the sophisticate, the self-improver. You may be positioned as being educational, or you could be entertaining and fun. You could be for the serious student or the casual family visitor. The tone of your communication could be serious and scholarly or aimed at fun. The point is to develop a distinct quality in your different types of communications and to be sure it is consistent with your educational mission, your collections, and your position in the marketplace. Once you have developed this quality in your communication in a way that reinforces the museum's mission while also attracting visitors, stay with it.

Be both specific and concise. When communicating with your publics through advertising, publicity, and literature, give the prospect an idea of what can be expected. What are the benefits? What are visitors going to get in exchange for their time and money?

Offer as much variety as possible. Consistent with the smart shopper and the value-conscious consumer, people today are looking for variety and diversity of activities. If variety is inconsistent with your established program, you may consider packaging with nearby attractions that provide experiences that augment your own.

Use a response vehicle in your advertising (and even in publicity when possible). Make sure there is a telephone number or address where people can call or write to get more information. But don't be discouraged if you don't get many inquiries from within a 150-mile radius. Experience shows that people who live close by simply don't feel the need for additional information prior to making their decision to visit. Responses will come primarily from those areas more distant from your location.

Train your on-site people. It is amazing how often people managing an institution are unaware that the behavior of their public-contact employees is driving away visitors. Likewise, these employees often are not aware of the impact that their behavior has on the visitor's ability to learn from and enjoy the museum. It is important to remember that visitors are in a "vacation" frame of mind, that they want to accomplish as much as possible in a limited amount of time, and that they want to avoid the hassles of the work day. They want to enjoy a relaxed and friendly atmosphere where they can contemplate the collections. Often they will want to share the enjoyable experience with their friends.

A CASE HISTORY

In recent years, Michigan's Henry Ford Museum and Greenfield Village has suffered an attendance decline that began in 1973. With the exception of the Bicentennial year, the decline continued at an accelerated rate through 1982. Eighty-three percent of the visitation loss occurred from 1979 to 1982, with the majority of the decline occurring in the summer months. Historically, 60 percent of attendance comes from the Great Lakes states where factory shut-downs and lay-offs caused unemployment to range upwards of 20 percent, the worst in the nation. Continuing loss of revenue caused the museum to dip into its endowment to meet expenses in 1980, 1981, and 1982. Knowing this practice held severe consequences if continued, the museum initiated the following program, creating a strategy designed to maximize return on investment.

In 1981, the museum established a curriculum committee to identify and evaluate the institution's strengths and to develop a new statement of mission. As a result of that committee's work, the following institutional mission was established: Henry Ford's Museum and Greenfield Village functions as a national museum of history and technology that tells the story of how America changed from a rural agrarian society to the urban industrial nation it is today.

In 1982, an extensive market research program was started to identify visitors and to determine how they perceived Henry Ford Museum and Greenfield Village. At the same time, the museum began collecting the zip codes of a random sample of visitors in order to accurately track their origin. Analysis of the research revealed the following key data:

> **Develop a distinct quality in your different types of communications and be sure it is consistent with your educational mission.**

Market Selection. Detroit was clearly the leading market and was key in maintaining a base of attendance. Markets between 150 and 250 miles from the museum contributed 21 percent of the total visitors, but accounted for 71 percent of the two-day ticket purchases. Therefore, these markets represented the greatest opportunity for return on marketing dollars

invested if attendance could be increased. Advertising was concentrated in Detroit plus six markets in the 150- to 250-mile radius.

Advertising Focus. While the museum was readily recognized in a majority of markets within a 250-mile radius, the prospects' image of the museum did not reflect the current institutional focus. It was necessary to redefine the museum complex in the marketplace to create a foundation upon which to build attendance in the coming years.

Timing. Seventy percent of all visitors were spending one or more nights away from home. Forty percent were spending three or more nights, and 20 percent were on vacations of seven or more days. This meant there was both an excursion and vacation target, and advertising was timed for the coincidence of both in order to build the summer season.

A POSITION THAT BREAKS TRADITION

Clearly, the museum's charter is to educate. The research showed, however, that while visitors felt that Henry Ford Museum and Greenfield Village was a great museum, they also thought the place was more fun to visit than most museums.

There were many reasons why visitors who were interviewed felt as they did. People can identify with many of the artifacts on a personal basis. Because some of these objects are designed in ways that seem strange by today's standards, they have a curious, even humorous, quality. Other reasons visitors gave included the person-to-person interpretation, alluring atmospheres such as a working historic farm and the home of famous Americans, and activities such as games and special event weekends.

The visitor's definition of the museum product as great and fun was consistent with the institution's overall mission. The position statement developed, "Henry Ford Museum and Greenfield Village, the Great American Museum That's Also Great Fun," broke the traditional museum mold. Advertising and publicity not only reflected this position, but were themselves suggestive of fun. Although one rarely sees the words "museum" and "fun" in the same sentence, there is no reason that the arts and the humanities can't be fun.

Large, high-impact newspaper ads were used to introduce the new position in the media in markets that produce the most visitors, and smaller space units were used to provide frequency and to develop a variety of messages. Sixty-second radio commercials ran in Detroit to aid in repositioning the museum to the local market. Multi-unit, small-space newspaper ads were also used in the local market to promote special event weekends.

Although the customers had already defined the institution as being "great" and "fun," the museum needed to establish a mechanism by which it could be assured that existing and new programs would be consistent with the new statement of mission and its new marketing position. This was accomplished by the formulation of an interpretation committee that was chaired by the director of education and was comprised of representatives from the curatorial, exhibitions, special programs, and marketing and public relations departments. The group evaluates every new program proposal and always takes the marketing position into account, along with the museum's overall mission, when making its recommendations.

Although its primary purpose is to be sure that all museum and village activities consistently address the overall mission, the interpretation committee has also been very successful in assuring that promises made in the marketplace are delivered on site. In fact, several new programs grew out of the visitors' desire for more involvement and the museum's desire to try new—and fun—ways to fulfill its mission.

Through a series of winter Great Escape Weekends inside the museum, the decades of the 1900s, '20s, '40s, and '50s are brought to life through such period activities as live music, dancing, foods, artifact demonstrations, and even personalities from the era as portrayed by interpreters. Also in the museum, a permanent activities center engages visitors in such diverse hands-on experiences as pedaling a highwheel bicycle and operating an assembly line.

In Greenfield Village, 19th-century games are taught and special event weekends have been refined to give visitors a greater degree of participation. For example, they may sign up and train with a Civil War unit during the annual Muzzle Loaders Festival.

As a result of these and other changes, attendance increased 13 percent over 1982, reversing a 10-year decline. Revenues for Henry Ford Museum and Greenfield Village increased 20.7 percent from 1982. In addition, the campaign won the 1983 Travel Industry Marketing Award as the best program in the nation for an attraction.

The campaign generated momentum as well as first-year results. In each of the next three years attendance has either met or exceeded goals, so that current ticket sales are running more than 20 percent ahead of 1983 levels—especially important at an institution where more than a third of the operating income is met by ticket sales, and a second third by selling food and merchandise to visitors.

The example of Henry Ford Museum and Greenfield Village demonstrates that museums can indeed compete for leisure time in the years ahead. But to be successful in the future will require a well-planned and thoughtfully executed program that brings the benefits of a visit to your facility into focus. Find your marketing niche, follow the philosophy of doing a few things well—and stay with the program as long as there is evidence that it is working. It will work to the considerable benefit of your museum and its visitors.

G. Donald Adams is director of public affairs for Henry Ford Museum and Greenfield Village. John Boatright, group vice president of the Martin Agency in Richmond, Virginia, was account supervisor for the Ford Museum campaign.

CASE STUDIES

Karen Johnson describes the transition from "working board" to "funding board" at the Discovery Museum of Orange County, explaining what was done and why.

Reflecting on the tremendous growth of Questacon, Michael Gore discusses the organizational strategy that transformed a prototype exhibition into Australia's national museum in just eight years.

FROM BIG IDEAS TO BIG BUCKS: THE BOARD IN TRANSITION *Karen Johnson*

You have a vision…a vision of kids learning about science in a fun, non-threatening way, in an attractive, busy, colorful place from knowledgeable people and with imaginative exhibits. At first you talk with your friends and business contacts about your vision, and before long you have created a board of directors. This first group of people, gathered almost at random, might consist of only eight to nine people with widely diversified backgrounds. But you share a common goal, so together you visit other science centers, talk with other experts and perhaps civic leaders, and ultimately draft a mission statement. Soon the specter of building or renovating a facility looms, and for that funding is required—there is no such thing as a "free building." The first place to seek funding is appropriately your board of directors.

Most new science centers are in the first or visionary stage. Board members are committed to the vision but often lack the financial means to accomplish the goal. At this point, the board needs to look critically at its composition. It is essential to begin enhancing the board membership in order to bring the science center into being. In looking for a fundraising board of directors, consider the three W's, namely Wealth,

Wisdom, and Work, as described by the former president of Brown University, Henry Wriston:

- Wealth—those with wealth and/or the ability to access sources of financial support for the institution

- Wisdom—those with expertise in specific areas, such as science, law, accounting, and marketing

- Work—volunteers who are willing to commit the time for necessary task-oriented functions

Each board member should fill a specific purpose and should know why he or she has been nominated. Ideally, each board member will be a balance of all three; however, two of the three should be mandatory. If a board member has only one W, his or her effectiveness will be severely limited.

Created in 1981, the Discovery Museum of Orange County is located on 11 acres of land, surrounded by industrial buildings, a park, and some residences. With four historic buildings which were moved onto the property in 1980 and 1982, the organization set out to restore the houses and start a living history museum for the 2.5 million people in the county. In 1984, a 1,500-square-foot science center located in a school about 10 miles away closed because the school district needed the classroom. The board of the science center merged with the board of the Discovery Museum three years later.

The museum staff at that time consisted of two full-time people (executive director and secretary) and four part-time people (program director and program leaders), and the facility was open three days per week. Motivated by a concern for the poor science achievement of students locally and nationally, the board embarked on an ambitious course with three goals:

1) to identify the parameters and costs of a science center appropriate to our population base

2) to enhance the board with individuals who could raise the significant capital costs of a new science center

3) to bring into the "family" persons with expertise and credibility in science and technology

The first goal was achieved through two feasibility studies. The last two goals involved the active participation of the existing board of directors. It was a formidable task, as in 1986 we had neither persons of wealth or visibility, nor anyone with scientific or technical expertise. Most of our board members were truly visionary. They understood the need to make the transition to a fundraising board, and knew this meant enlarging their number with persons from a different socio-economic level. They rewrote the bylaws to allow the board to expand, eliminating the need to "kick anyone out" immediately, and decided to risk changing the make-up of the board. As it turned out, we were able to add the kind of new people we needed and still get along.

We began by identifying any and all contacts our board members and small staff had. Several years before, the CEO of the largest corporation headquartered in Orange County had been asked to help with the living history museum. (He had played in one of the houses as a child.) He invited 24 prominent persons to have breakfast in the dilapidated house; nine responded positively, and five showed up. Of those five, one sent a note about four months later, thanking us for the breakfast and indicating that her foundation would not be able to give us any funding that year—an amazing comment, as we had not asked for any funding. So I called and arranged a lunch with her, the board president (president of a local Cadillac dealership), the then executive director, and myself, the part-time development director.

The woman we were approaching had been in the oil business in Indonesia with her husband for 40 years, but upon his death had returned to Orange County. She accepted the luncheon invitation. We planned strategy for days. We had four different requests, beginning with asking her to join the board of directors. Lunch was agony. The board president, an extremely nice person, was many years this woman's junior. He talked about everything except our project. Finally, two hours later, after the check was paid and we were about to stand up and leave, he offhandedly asked, "Would you be interested in joining the board of directors?" Graciously, she said yes, and we had our first board member with all three W's.

Next, the board believed that a county supervisor would add visibility and stature to our fledgling organization. We selected the supervisor for the district in which the living history museum was located. He possessed two W's, wisdom and access to wealth in the form of available county money. He was very careful and very thorough. He asked for articles of incorporation, bylaws, evidence of nonprofit status, and

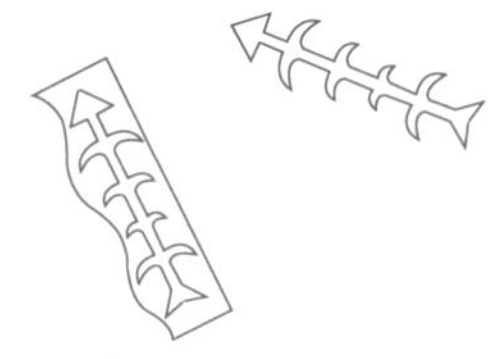

financial statements. He wanted to know who else was involved. Finally, after we had met several times with his administrative assistant, the supervisor agreed to meet with us for lunch. We had a more relaxed lunch this time and recruited a two-W, highly visible board member.

The next was almost as easy. Son of a major residential developer in Orange County, and owner of a very successful development business, he was a close business associate of our first new board member. At our suggestion, she wrote him a letter, describing her involvement (four years at that point) with the Discovery Museum. I followed up with several telephone calls and, through his secretary, set up a time for him to come out to our present site and talk with me about our future plans. After his visit, in which he voiced his concern that we weren't credible in science, he agreed to join the board. We had our second three-W board member.

As we made progress on feasibility studies, our first three-W board member suggested we meet with the associate president of a private local college. He had a Ph.D. in chemistry and from our first meeting was an outspoken advocate of the science center. He agreed to join the board, a strong two-W, with expertise in the sciences.

The largest aerospace company in southern California was the source of our next two three-W board members. We had been cultivating a relationship with this company through our history program with the specific intention of introducing the science project at the appropriate time. When we began our process of board enhancement, we asked that one of their senior level executives join the board. He did. Two years later, he in turn recruited the highest ranking company officer into the fold. These two individuals have been ideal examples of the three W's. They have access to other high-ranking corporate officers and meet with them regularly to ask for capital gifts (they work). Because of their professional backgrounds, they knew in their gut how important science centers are to increasing science literacy (they have the wisdom to sell the project and guide programming). And they have corporate clout, thus they were able to persuade their own company officers to make the largest gift to our campaign (they have access to wealth).

Science centers are not built through the dreams and wishes of one or two people.

In Orange County, we have a Business Committee for the Arts (BCA). This organization was started by CEOs and other top corporate leaders to assist arts and cultural organizations. The executive director matches requests for board members with people in the business community who show an interest in a specific project. One of the most recent board members came to us through this source. He is an M.D. who started his own molecular biology company and currently is managing partner of one of the largest venture capital companies locally. I met with him and the executive director of BCA, and by now, we had both feasibility studies and some additional public relations materials with colored photographs and videotapes of kids and exhibits at other science centers. This man was a knowledgeable advocate of science centers because of his professional background and experience at other science centers. With three small children, he was very interested in having one available in Orange County. Chalk up another three-W!

Science centers are not built through the dreams and wishes of one or two people. Strong community support with funding and planning through volunteer efforts is needed. And a board of directors who will meet with people, talk about the project, and give generously of their time and money is vital. It doesn't hurt to have an astronaut either.

Unquestionably the most recognized member of our board of directors is Buzz Aldrin. The second astronaut to step on the moon has long been interested in improving science education. We were just lucky that he moved to Orange County and while at a social engagement, met the current president of our board of directors. With no planning, but great enthusiasm, he asked Buzz and his wife to join our board of directors. They accepted! Aldrin's participation at board meetings has been limited, but when there, he has good ideas and is an enthusiastic participant. Sometimes good luck is better than good planning!

Because Hispanics represent 90 percent of the students in Santa Ana, and are the largest minority population in Orange County, we were delighted when the mayor of Santa Ana, who is Hispanic, volunteered to help the Discovery Museum. Through his contacts, we have met with and will be working with the president of a marketing and public relations company that specializes in advertising for national products in the Hispanic markets. It may take more effort to identify diverse board members, but they often see the need more keenly and bring excellent resources.

When changing the board from a working to a fundraising body, you must let the old guard know how valuable they are, but that what were once critical agenda items are now going to take a back seat to raising significant capital. This focus on raising funds must be unwavering. When working board members persist in discussing operating issues at the expense of discussing campaign strategies, it may be time to divide the group by conducting two meetings monthly—a board meeting and a campaign cabinet meeting. In this way, the old guard board can focus on the more general concerns of the organization, while the three-W's can work on fundraising.

Bringing the right mix of people together to form the backbone of support and add validity to a new or developing science center is a mixture of hard work, creative planning, luck, and making the request. There are so many people who have the time, money, and interest to help, and you just need to ask. Success breeds success: We now have a bank president, the CEO of an international pharmaceutical corporation, and the local president of the largest regional newspaper as well, and our board is now accustomed to high achievement. As a postscript, let me note that $3.3 million has been pledged by the new board members mentioned above. You may not have 100 percent success, but building a new board is an exciting and intriguing process.

Karen Johnson is executive director of Discovery Museum of Orange County, which includes the Discovery Science Center and a living history center. She has a B.A. in chemistry. Ms. Johnson has served as a board member of Discovery Museum and a number of other nonprofit organizations, and as president of the San Francisco Junior League.

QUESTACON: STRATEGIES FOR EXPANSION *Michael Gore*

Questacon, the first interactive science center in Australia, established in 1980, was inspired, as so often has been the case, by the Exploratorium in San Francisco. In only eight years, it developed from a modest prototype housed in an old school to a national institution in a specially designed building in the heart of Australia's capital city—a rate of growth remarkable even among science centers.

Questacon was conceived on the campus of the Australian National University in Canberra. The university gave the project respectability, although not much else, and enabled it to attract the financial support it required for survival from the public, commerce, and industry.

One of the first problems Questacon faced was the fact that it was situated in Canberra. Whereas Canberra is the capital of Australia, it is devoid of any major industrial corporations, whose headquarters are in Sydney or Melbourne. Canberra is perceived by the corporate sector as a civil service city; corporations tend to feel that the federal government should take care of anything that is built and run in the national capital.

There was also the view that a science center with aspirations of serving the nation should not be built in Canberra at all. The current population of Australia is 17 million, with Sydney having 3.7 million people, Melbourne 3.1 million, and Canberra 300,000.

Notwithstanding the problem of its geography, Questacon led

a charmed life in its early days. It was housed in an old school building for a peppercorn rent, and cleaning, heating, lighting, power, local telephone calls, and maintenance were all included. This made the task of making ends meet relatively easy. Within the first two years, Questacon became financially self-sufficient, obtaining modest donations from companies like Shell and Esso, and from community grants programs within Canberra; and additional revenue from admission fees and shop sales. We also received several donations for specific exhibits. Questacon developed steadily until, by 1987, it contained some 150 interactive exhibits and employed approximately 150 explainers, ranging in age from 18 to 80. We believe that explainers play a major role in the success of interactive science centers and the decision was taken to provide them some form of remuneration right at the start of the project.

To operate in this manner in the early 1980s was a big departure from the norm in Australia. Unlike the situation in the United States, entry to nearly all Australian museums and art galleries at that time was free, and they were completely funded by either state or federal governments. In contrast, Questacon began life fighting for survival in exactly the same way as science centers in the United States.

The transformation of the prototype Questacon into a major national institution involved a shift in ownership from the Australian National University to the federal government. The government inherited an organization that had demonstrated that it was able to survive without the government providing all funds to meet recurrent costs. This coincided with a major government rethink about its attitudes toward museums throughout the country. Since 1988, when Questacon became Questacon—The National Science and Technology Centre, many museums throughout Australia have reluctantly introduced admission charges.

In 1984, the Australian Bicentennial Authority set up a working party to seek ideas in the area of science and technology which could be included in the Australian Bicentennial celebrations of 1988. Four members of the working party were based in Canberra and were familiar with the excitement and interest generated by the fledgling Questacon. After long debate and bureaucratic wrangling, the working party recommended that a national science center be established, with Questacon acting as a prototype. When the decision was taken in 1986 to go ahead with the project, Questacon had been operating for six years and was acting as the model for other, smaller interactive science centers that were, by this time, springing up all over Australia.

When it came to the problem of finding funds to cover the capital cost of the large new building, once again the project appeared to be leading a charmed life. With the federal government's decision to go ahead, there came a simultaneous decision from the Japanese government to provide half of the capital cost of construction of the new center on behalf of the people of Japan and the Japanese business community. The rest came mostly from the Australian federal government.

The transformation from Questacon-the-prototype to Questacon—The National Science and Technology Centre also meant changing from academia to civil service. We started with no paid full-time staff, and now have about 60 permanent staff who are civil servants, and about 280 part-time explainers. As the center expanded, the administrative positions were filled by civil servants; the design and construction team by scientists, graphic designers, and technicians; and the education section primarily by science teachers. All staff have to be employed under the terms and conditions of the Australian Public (Civil) Service. Questacon does not have the ability to hire and fire at will. All posts have a carefully written set of selection criteria and a comprehensive duty statement. Thus the strategies we used to "people" the new place were prescribed for us. The main problem in the hiring of staff was the length of time it took—sometimes upwards of six months.

In accepting responsibility to establish and to provide a measure of recurrent funding for the national science center, the federal government made no ongoing commitment to major exhibitions and educational programs, expecting the center to fund them from private sector sponsorship. The government's annual budget appropriation provides for such items as permanent salaries, building-running costs, repairs and maintenance, and all the day-to-day administrative costs of the center—about half of our annual operating budget of $6 million. (All figures in this article are quoted in Australian dollars; $1 Australian is worth about 70 cents U.S.) The other $3 million must be generated by the center.

In only eight years, Questacon developed from a modest prototype to a national institution—a rate of growth remarkable even among science centers.

As in Questacon's earlier phase, we decided to use the same strategies as North American science centers to generate the $3 million in operating income not covered by the government. But we also introduced a new ingredient: taking the word "national" in our title very seriously. Here is how the funds are generated:

- Annual visitation is about 350,000. There is an admission charge ($6 for adults, $3 for children, and a $45 family membership rate) which yields about 45 percent of the annual earned income. In addition, the center has a retail store, and the building is hired out for private functions.

- Operating a catering facility can be a vexed question for science centers. Our experience with outside caterers was not good, and the center now runs its own cafeteria, which is making a significant contribution to the center's revenue.

- If, during the course of a financial year, it becomes clear that the income from these various revenue sources will not achieve the required target, then various expenditure-cutting strategies are set in train. Building maintenance programs are postponed, staff who resign are not immediately replaced, and in some, but rare, cases, educational programs are clipped.

- We seek support from other federal government programs. For example, we have obtained funds from the Department of Employment, Education, and Training to take hands-on science to the Aboriginal children in remote areas of the Australian outback.

- When it comes to funding a major exhibition, we solicit support by telling sponsors that any major national exhibition will not only run for 12 or 15 months in the national capital, but will then be toured to other science centers in Australia. On tour, it will bear both the name of Questacon—The National Science and Technology Centre, and the name of the sponsor. In this way both get

nationwide recognition. In the last three years, the center has attracted nearly $5 million in sponsorship for its exhibitions and traveling programs. Currently, with exhibitions not only in Canberra, but also at centers in Melbourne, Sydney, and Adelaide, we are fulfilling our aim of being a truly national institution.

In order to maintain their popularity—and admissions income—interactive science centers must regularly change exhibitions. Questacon contains five galleries, each about 5,000 square feet. The strategy that has so far been successfully adhered to is to have a new exhibition in one gallery every 12 to 18 months. Continually rolling over exhibitions is not easy and places the center's design and construction staff under perpetual pressure. In order to relieve the pressure, we are canvassing for collaboration between Australian science centers, eventually including the newly emerging New Zealand science centers. Perhaps we will eventually embrace all the members of the Association of South East Asian Nations. We wish to form cartels between several new centers with the purpose of jointly designing and constructing new exhibitions.

Our strategy of "going national" has also proved extremely successful in promoting Questacon and its programs to the Australian public. Rather than simply operating in Canberra, a number of programs travel all over the continent, which, I wish to point out, is almost exactly the same size as the continental United States. The center runs the Shell Questacon Science Circus, which was modeled after the Ontario Science Circus, but with an interesting new component. The Shell Questacon Science Circus is operated by 10 specially selected Australian science graduates who are enrolled in a post-graduate diploma course in Science Communication at the Australian National University. They are trained to communicate science to the public via media such as radio, television, and public speaking, and they practice their craft on the general public throughout the length and breadth of the land with the Science Circus.

Questacon also runs traveling planetariums (the inflatable variety made by Learning Technologies, Inc., in Massachusetts) and national science lecture programs, including Questacon Science Theatre, a program of two-and-a-half-hour evening performances of science demonstrations for the general public. Tickets sold for admission to the science theater also give half-price entry to the center in Canberra. All of these educational programs have the dual role of raising revenue for the center, and at the same time, raising its national profile.

Our current success with programs that "go national" defines one of the strategies that the center must adopt in planning continued growth. We expect that by the year 2000, Questacon—The National Science and Technology Centre will devote about 75 percent of its resources to outreach programs and 25 percent to the exhibitions and activities in Canberra. It must not be inferred from these target figures that there is any intention to short-change the center in Canberra. What is planned is to find new ways, means, and resources to expand the number of traveling programs, while still ringing the changes on the displays and activities in Canberra.

It may well be that in order to achieve these goals, a subsidiary company or organization has to be formed. This may not be without problems, for it will require building an interface between the civil service and a commercially-oriented organization. We may need to invent new strategies to further develop Questacon as an effective and efficient national resource for the promotion of science to the Australian public.

Born and educated in the north of England, Michael Gore began teaching physics in 1962 at the newly created Australian National University in Canberra. En route for Europe in 1975, he discovered the Exploratorium, and in 1977, began work to establish Australia's first interactive science center. For his involvement with the development of Questacon, and his services to science education, Mr. Gore was made a member of The Order of Australia in 1983. In 1986, he became the foundation director of Questacon—The National Science and Technology Centre.

APPENDIX A
A MESSAGE TO SCIENCE CENTERS

Based on two surveys of ASTC member science centers, one in 1986 and a second in 1990, the ASTC Board of Directors makes the following specific recommendations to museum directors and staff. These recommendations reflect the priority our members place on improving service to underserved audiences: minorities, women, and people with disabilities.

FOR MUSEUM DIRECTORS:

1. *Strengthen ties to key leaders in the underserved community you seek to reach.* Seek out elected officials, religious leaders, community activists, university faculty. Acquaint them with your institution and tell them of your goals. Learn how your museum is perceived in their community, and ask their advice about how to increase visitors.

2. *Develop an advisory board comprised of community leaders and your own board members* to help you develop an action plan, to become ambassadors for the museum, and to provide you with a sounding board. Use this as an interim measure while working to get these leaders on your board.

3. *Conduct your own informal surveys of how your museum is serving a newly targeted audience.* Count how many minority, female, and disabled visitors the museum attracts on any given day or time. See if they are coming in school groups, or as members of the general public. Ask minority, female, and disabled visitors what they like and don't like about the museum. Observe how they are treated by floor staff, security guards, and restaurant staff.

4. *Convene professional development seminars on understanding and communicating across different cultures.* Many corporations have instituted cultural diversity training to boost productivity and increase customer bases. You may find one willing to donate the services of an organization development expert. Circulate articles and reading lists, and set aside informal discussion time for staff.

5. *Tell your staff that diversifying the museum's audience is an important priority.* Let the floor staff know that welcoming all visitors is a critical part of their job, one that will be included in their performance review. Make progress reports a regular part of senior management meetings. Ask all departments to identify actions they will take as part of their annual plans. With every new project, ask the staff to identify how equity concerns can be addressed in the process.

6. *Step up efforts to recruit and train minority, female, and disabled staff members, particularly for professional level positions.* An integrated staff is a signal, particularly to minority audiences and people with disabilities, that the museum is a safe place to come and to bring their children. An integrated staff increases the flow of ideas across cultures, and helps you find sensible, effective ways to incorporate various cultural perspectives. An integrated staff also provides role models for young people.

7. *Recruit more board members who are minority, female, or people with disabilities.* Most of our institutions are located in urban settings that are becoming increasingly diverse. Our communities need to see themselves reflected in our *leadership* and we, in turn, need access to a broader constituency to prosper and grow.

8. *Target financial resources to expanding your audience.* Any truly important institutional priority shows up in the budget. Consider this a long-range investment which, like any other, is tracked carefully by the senior management, and expected to yield returns.

FOR MUSEUM STAFF:

1. *Notice who comes to the museum.* Count how many minority, female, and disabled visitors the museum attracts on any given day or time. See if they are coming in school groups, or as members of the general public. Ask minority, female, and disabled visitors what they like and don't like about the museum.

2. *Notice your behavior toward visitors.* Floor staff can notice whether they are more helpful to some visitors than others. Demonstrators and educators can notice whether they invite involvement more frequently from one particular kind of visitor than from others. Exhibit designers can notice who lingers to use an exhibit, and what minority, female, and disabled visitors say about it.

3. *Learn about other communities.* Expand your reading list to include popular publications targeting an underserved audience that your museum wants to reach. Find out whether local universities have scholars with relevant specialties, particularly in their history departments. Call ASTC for suggestions of publications.

4. *Look for ways to increase the flow of information between the museum and the community it seeks to serve.* Our daily professional lives include many opportunities to gather information, seek advice, or promote a new endeavor. Remember your role as an ambassador of the museum, and in particular, listen when someone from the targeted audience is talking.

5. *Reflect the institution's concern in dealings with your staff.* Support your director's efforts to provide leadership in this area. Follow through on any institution-wide steps to address cultural awareness concerns. Make equity issues a part of performance reviews, action plans, project designs, or any other management tools at your disposal.

PROFILES OF TYPICAL SCIENCE CENTERS

I. VERY SMALL SCIENCE CENTER

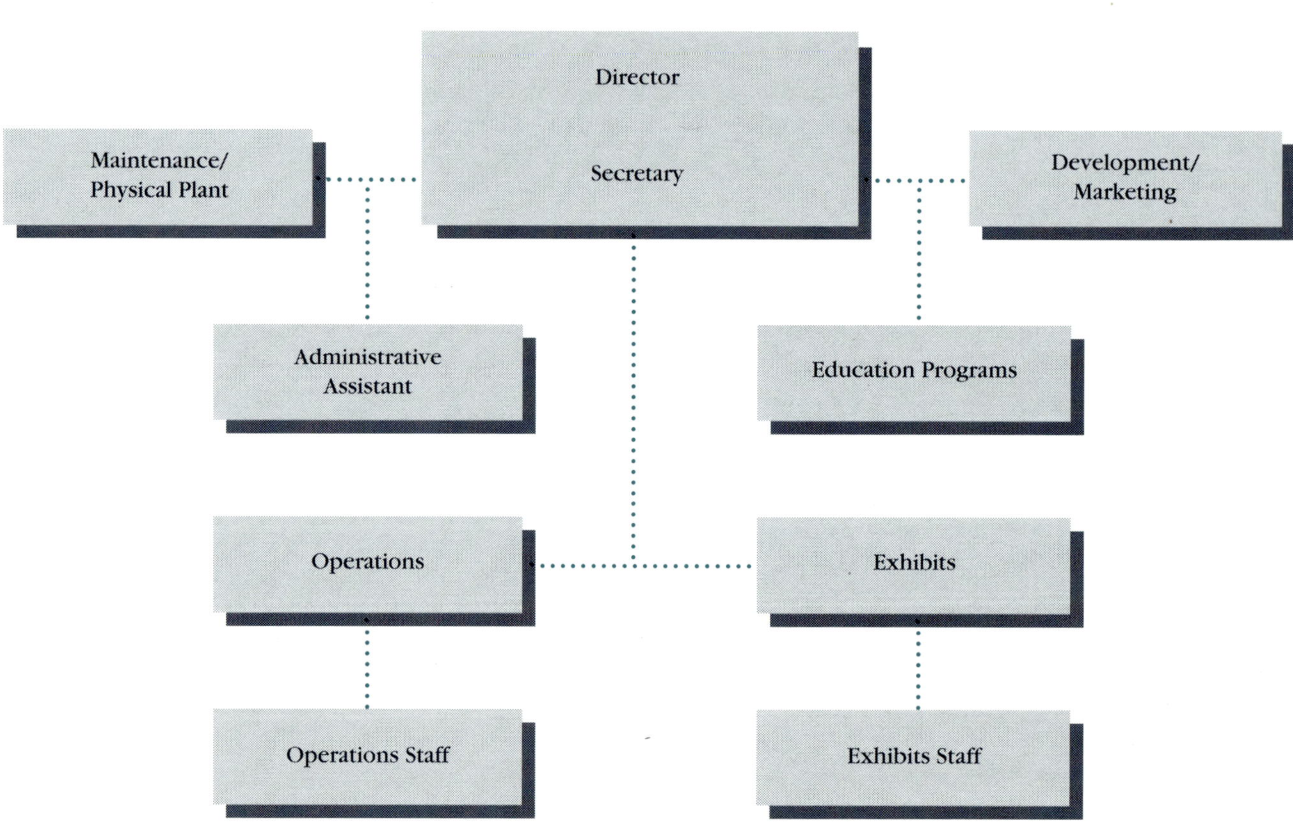

STAFF	FTE	VOLUNTEERS	
		N	*FTE*
Administration/Finance	2.20	4	.6
Development/PR	1.00	4	.5
Operations	2.00	11	.9
Plant	1.70	3	.3
Exhibits	1.60	6	.7
Planetarium/Theater	1.75	3	.7
Programs/Education	2.80	10	.9
Museum Store	1.00	10	.7
Collections/Research	1.50	3	.7
Other	1.50	45	5.1
Total	17.05	99	11.1

ANNUAL ATTENDANCE

78,000 visitors

USE OF SPACE	SQ. FT.	%
Exhibits	5,200	45
Planetarium/Theater	990	9
Food Service & Store	360	3
Education	1,200	10
Other Public Space	680	6
Non Public Space	3,100	27
Total	11,530	100

OPERATING BUDGET

INCOME	$	%
Revenue		
Admissions	75,800	39
Investments	4,650	2
Membership	19,800	10
Store	18,100	9
Food Service	135	0
Special Events	19,200	10
Programs	23,000	12
Other	34,800	18
Total Revenue	$195,485	100
Support		
Individuals	40,200	9
Foundations	37,300	9
Corporations	36,800	9
Federal Government	87,500	20
State Government	29,500	7
Local Government	89,800	21
Other	111,200	25
Total Support	$432,300	100
Total Income	$627,785	

EXPENSES	$	%
Exhibits/Collections	55,900	9
Education	66,200	11
Planetarium/Theater	25,500	4
Membership	13,800	2
Other Programs	13,700	2
Management/General	107,000	17
Plant	70,900	11
Development/PR	31,500	5
Store	17,700	3
Auxiliary Enterprises	16,700	3
Other Support Services	199,000	32
Total Expenses	$617,900	100
Surplus (Deficit)	$9,885	

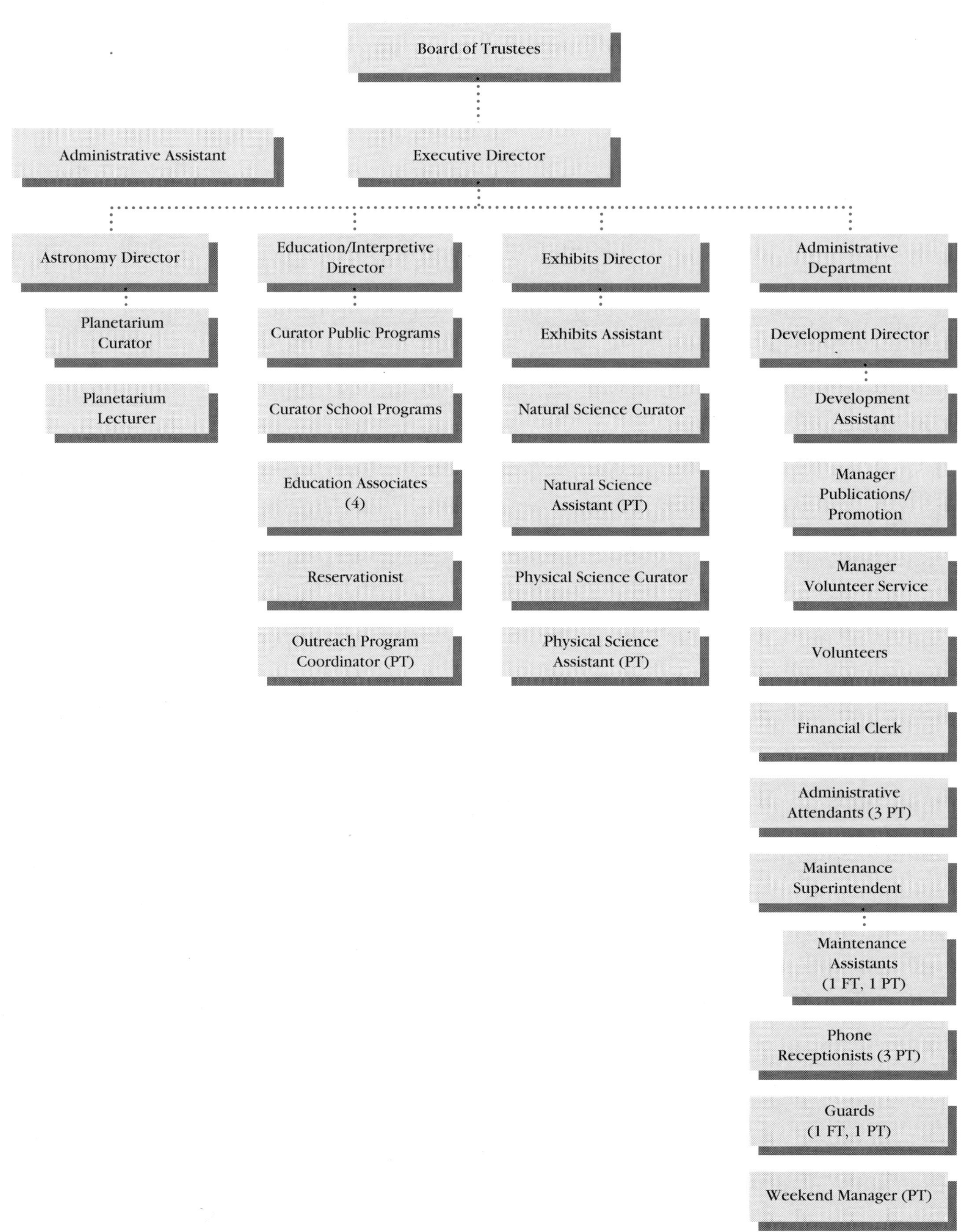

Board of Trustees
Administrative Assistant
Executive Director
Astronomy Director
Education/Interpretive Director
Exhibits Director
Administrative Department
Planetarium Curator
Curator Public Programs
Exhibits Assistant
Development Director
Planetarium Lecturer
Curator School Programs
Natural Science Curator
Development Assistant
Education Associates (4)
Natural Science Assistant (PT)
Manager Publications/ Promotion
Reservationist
Physical Science Curator
Manager Volunteer Service
Outreach Program Coordinator (PT)
Physical Science Assistant (PT)
Volunteers
Financial Clerk
Administrative Attendants (3 PT)
Maintenance Superintendent
Maintenance Assistants (1 FT, 1 PT)
Phone Receptionists (3 PT)
Guards (1 FT, 1 PT)
Weekend Manager (PT)

STAFF	FTE	VOLUNTEERS	
		N	*FTE*
Administration/Finance	5.20	11	.6
Development/PR	3.00	26	2.0
Operations	6.20	12	1.5
Plant	4.50	6	.8
Exhibits	4.20	7	1.2
Planetarium/Theater	3.40	9	.9
Programs/Education	6.60	31	4.3
Museum Store	2.60	16	2.2
Collections/Research	3.90	10	1.4
Other	3.30	39	2.6
Total	42.90	167	17.5

ANNUAL ATTENDANCE

189,000 visitors

USE OF SPACE	SQ. FT.	%
Exhibits	16,700	39
Planetarium/Theater	3,500	8
Food Service & Store	1,000	2
Education	3,400	8
Other Public Space	3,750	9
Non Public Space	14,900	34
Total	43,250	100

OPERATING BUDGET

INCOME	$	%
Revenue		
Admissions	353,700	37
Investments	119,200	13
Membership	94,700	10
Store	145,700	15
Food Service	24,700	3
Special Events	89,600	9
Programs	78,400	8
Other	41,100	4
Total Revenue	$947,100	100
Support		
Individuals	272,000	22
Foundations	139,500	11
Corporations	67,300	5
Federal Government	66,900	5
State Government	116,300	9
Local Government	367,000	29
Other	215,300	17
Total Support	$1,244,300	100
Total Income	$2,191,400	

EXPENSES	$	%
Exhibits/Collections	263,100	14
Education	192,800	10
Planetarium/Theater	79,700	4
Membership	65,700	4
Other Programs	202,300	11
Management/General	271,000	15
Plant	263,900	14
Development/PR	146,700	8
Store	147,700	8
Auxiliary Enterprises	70,500	4
Other Support Services	152,100	8
Total Expenses	$1,855,500	100
Surplus (Deficit)	$335,900	

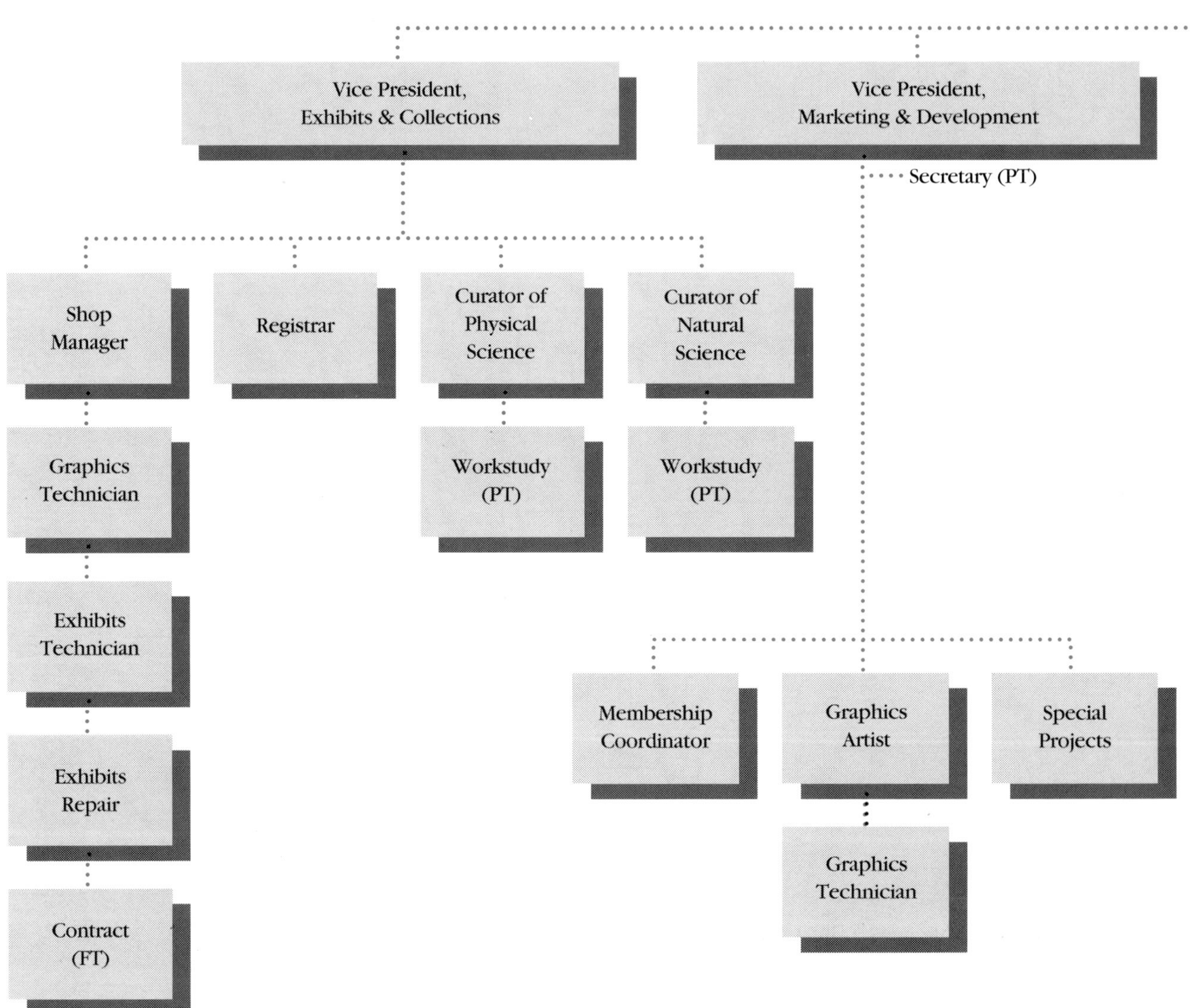

Vice President, Exhibits & Collections
Vice President, Marketing & Development
Secretary (PT)
Shop Manager
Registrar
Curator of Physical Science
Curator of Natural Science
Graphics Technician
Workstudy (PT)
Workstudy (PT)
Exhibits Technician
Exhibits Repair
Contract (FT)
Membership Coordinator
Graphics Artist
Special Projects
Graphics Technician

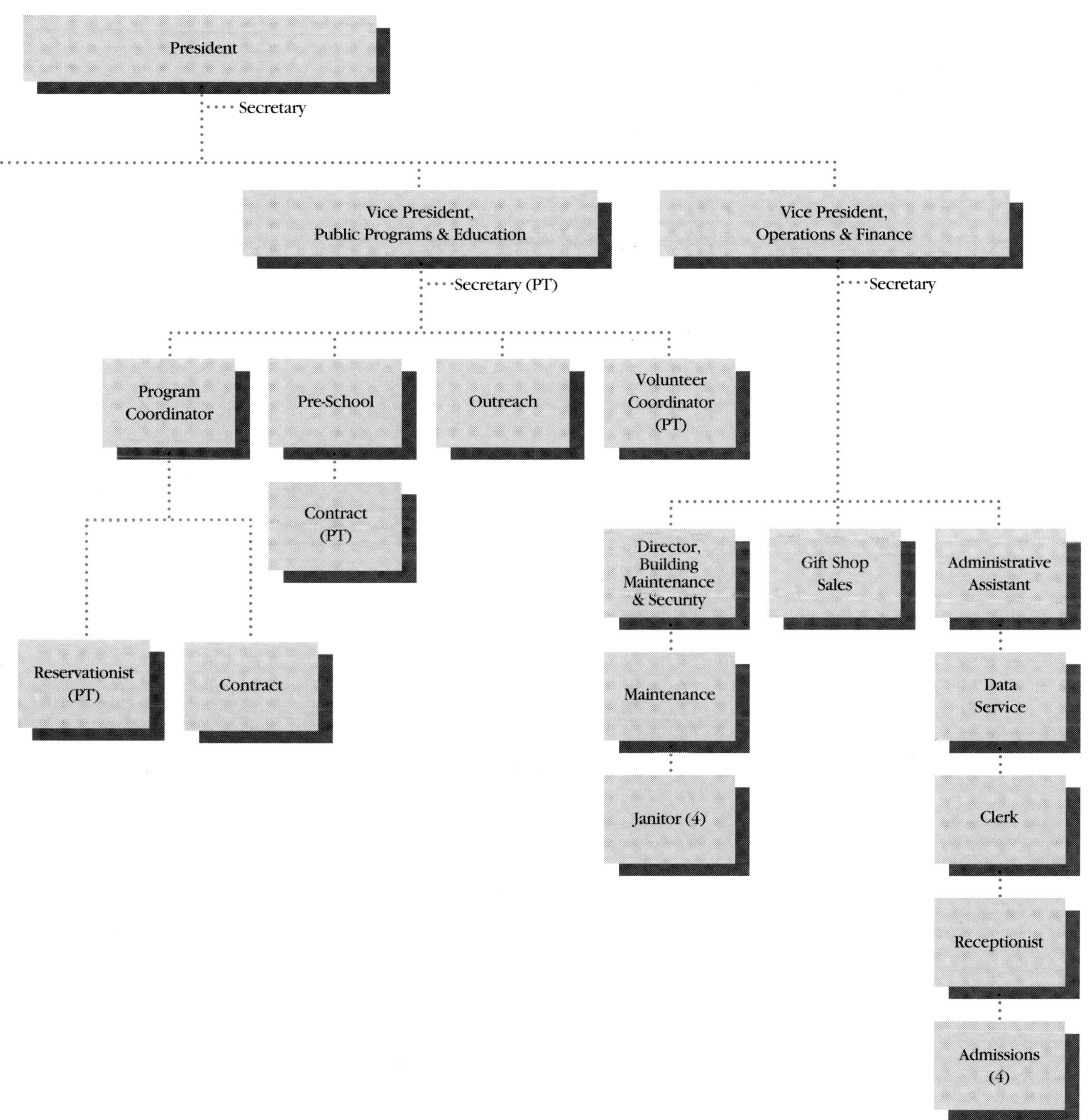

President
Secretary
Vice President, Public Programs & Education
Vice President, Operations & Finance
Secretary (PT)
Secretary
Program Coordinator
Pre-School
Outreach
Volunteer Coordinator (PT)
Contract (PT)
Director, Building Maintenance & Security
Gift Shop Sales
Administrative Assistant
Reservationist (PT)
Contract
Maintenance
Data Service
Janitor (4)
Clerk
Receptionist
Admissions (4)

STAFF	FTE	VOLUNTEERS	
		N	*FTE*
Administration/Finance	10.40	15	1.2
Development/PR	7.10	25	1.8
Operations	15.20	26	2.3
Plant	10.10	24	.9
Exhibits	10.70	19	2.6
Planetarium/Theater	5.90	3	.4
Programs/Education	16.70	144	9.6
Museum Store	4.70	15	1.4
Collections Research	14.90	20	2.5
Other	14.30	13	1.2
Total	110.00	304	23.9

ANNUAL ATTENDANCE 407,000 visitors

USE OF SPACE	SQ. FT.	%
Exhibits	47,800	37
Planetarium/Theater	7,800	6
Food Service & Store	3,600	3
Education	7,600	6
Other Public Space	11,900	9
Non Public Space	52,200	40
Total	130,900	100

OPERATING BUDGET

INCOME	$	%
Revenue		
Admissions	539,200	24
Investments	154,600	7
Membership	178,600	8
Store	264,200	12
Food Service	96,800	4
Special Events	215,900	10
Programs	373,300	17
Other	400,500	18
Total Revenue	$2,223,100	100
Support		
Individuals	193,800	7
Foundations	315,900	11
Corporations	280,100	9
Federal Government	269,400	9
State Government	739,200	25
Local Government	918,100	31
Other	253,700	9
Total Support	$2,970,200	100
Total Income	$5,193,300	

EXPENSES	$	%
Exhibits/Collections	572,700	12
Education	685,300	14
Planetarium/Theater	278,000	6
Membership	160,000	3
Other Programs	818,700	17
Management/General	602,700	13
Plant	543,300	11
Development/PR	222,300	5
Store	336,800	7
Auxiliary Enterprises	103,200	2
Other Support Services	489,400	10
Total Expenses	$4,812,400	100
Surplus (Deficit)	$380,900	

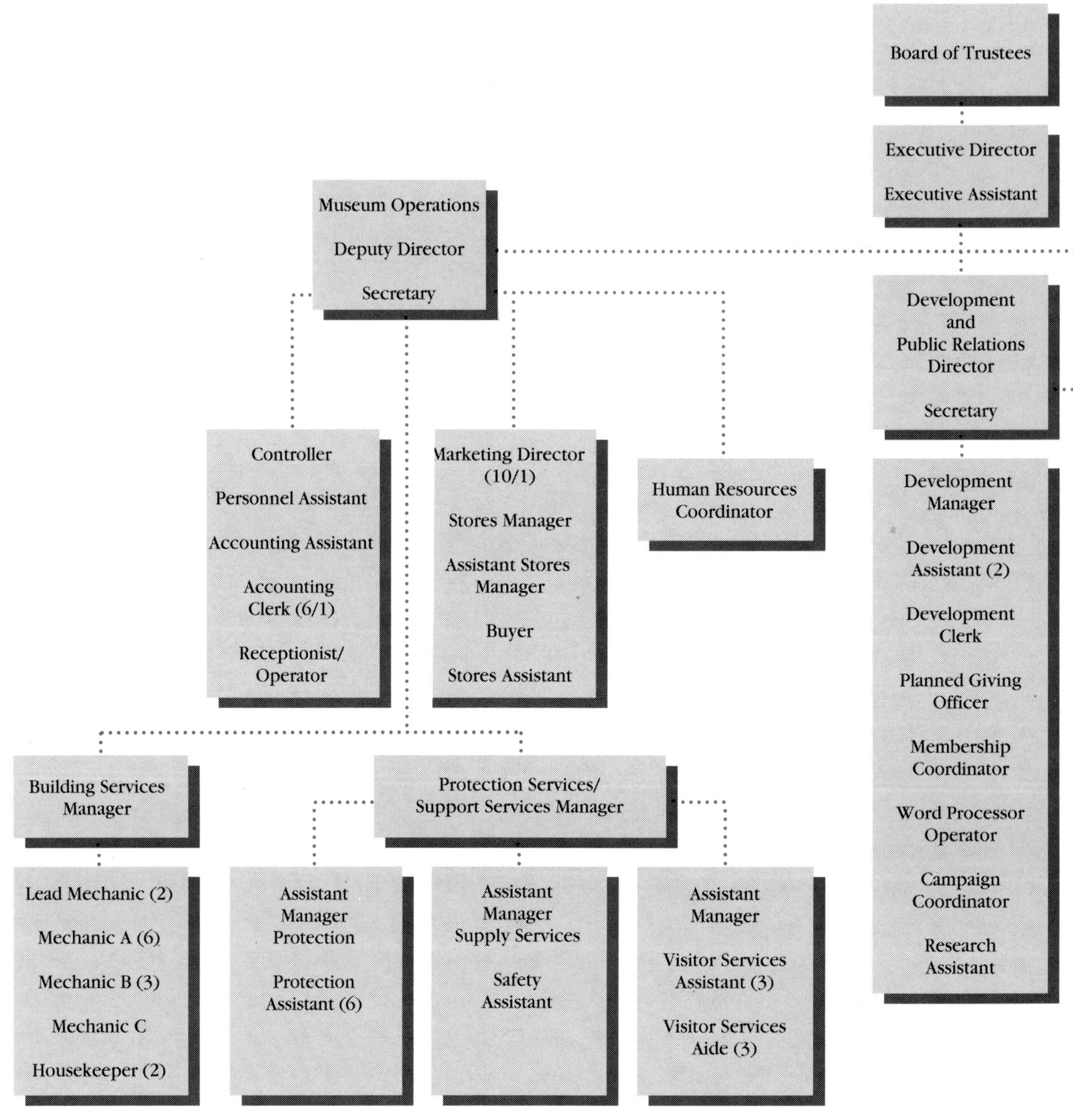

Board of Trustees
Executive Director
Executive Assistant
Museum Operations
Deputy Director
Secretary
Development
and
Public Relations
Director
Secretary
Controller
Personnel Assistant
Accounting Assistant
Accounting
Clerk (6/1)
Receptionist/
Operator
Marketing Director
(10/1)
Stores Manager
Assistant Stores
Manager
Buyer
Stores Assistant
Human Resources
Coordinator
Development
Manager
Development
Assistant (2)
Development
Clerk
Planned Giving
Officer
Membership
Coordinator
Word Processor
Operator
Campaign
Coordinator
Research
Assistant
Building Services
Manager
Protection Services/
Support Services Manager
Lead Mechanic (2)
Mechanic A (6)
Mechanic B (3)
Mechanic C
Housekeeper (2)
Assistant
Manager
Protection
Protection
Assistant (6)
Assistant
Manager
Supply Services
Safety
Assistant
Assistant
Manager
Visitor Services
Assistant (3)
Visitor Services
Aide (3)

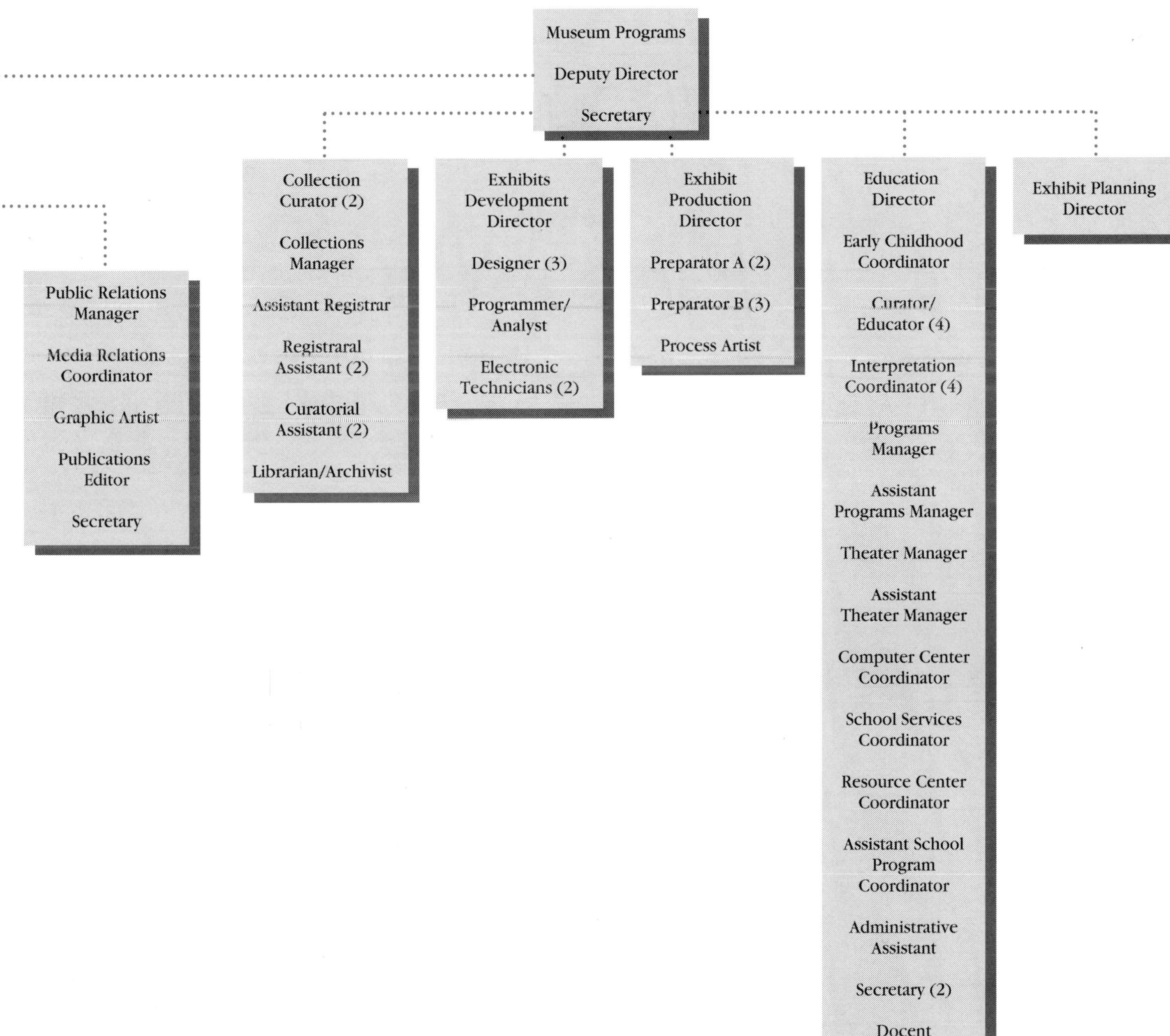

Museum Programs
Deputy Director
Secretary

Public Relations
Manager

Media Relations
Coordinator

Graphic Artist

Publications
Editor

Secretary

Collection
Curator (2)

Collections
Manager

Assistant Registrar

Registraral
Assistant (2)

Curatorial
Assistant (2)

Librarian/Archivist

Exhibits
Development
Director

Designer (3)

Programmer/
Analyst

Electronic
Technicians (2)

Exhibit
Production
Director

Preparator A (2)

Preparator B (3)

Process Artist

Education
Director

Early Childhood
Coordinator

Curator/
Educator (4)

Interpretation
Coordinator (4)

Programs
Manager

Assistant
Programs Manager

Theater Manager

Assistant
Theater Manager

Computer Center
Coordinator

School Services
Coordinator

Resource Center
Coordinator

Assistant School
Program
Coordinator

Administrative
Assistant

Secretary (2)

Docent

Exhibit Planning
Director

STAFF	FTE	VOLUNTEERS	
		N	*FTE*
Administration/Finance	20.00	24	1.10
Development/PR	10.40	10	.70
Operations	63.70	3	.20
Plant	25.10	0	.00
Exhibits	35.00	5	.40
Theater/Planetarium	6.70	10	1.25
Programs/Education	19.60	214	11.00
Museum Store	7.90	16	1.60
Collections Research	32.20	19	2.00
Other	16.30	60	24.00
Total	236.90	361	42.25

ANNUAL ATTENDANCE

1,546,000 visitors

USE OF SPACE	SQ. FT.	%
Exhibits	198,500	45
Planetarium/Theater	15,800	4
Food Service & Store	12,100	3
Education	11,300	3
Other Public Space	62,900	14
Non Public Space	137,200	31
Total	437,800	100

OPERATING BUDGET

INCOME	$	%
Revenue		
Admissions	1,436,200	29
Investments	803,300	16
Membership	241,600	5
Store	692,300	14
Food Service	878,200	18
Special Events	291,600	6
Programs	312,500	6
Other	240,200	5
Total Revenue	$4,895,900	100
Support		
Individuals	307,900	4
Foundations	135,500	2
Corporations	238,900	3
Federal Government	50,000	1
State Government	3,219,200	46
Local Government	1,897,000	27
Other	1,205,900	17
Total Support	$7,054,400	100
Total Income	$11,950,300	

EXPENSES	$	%
Exhibits/Collections	1,651,400	14
Education	1,109,500	10
Planetarium/Theater	732,200	6
Membership	289,500	3
Other Programs	667,000	6
Management/General	1,063,600	9
Plant	2,704,800	24
Development/PR	498,500	4
Store	696,800	6
Auxiliary Enterprises	1,040,000	9
Other Support Services	1,019,900	9
Total Expenses	$11,473,200	100
Surplus (Deficit)	$477,100	

Reprinted from Susan McCormick, ed., *The ASTC Science Center Survey: Administration and Finance Report.*

BIBLIOGRAPHY

Adams, G. Donald. *Museum Public Relations*. Nashville, TN: American Association for State and Local History, 1983.

Alt, Michael B. "Four Years of Visitor Surveys at the British Museum (Natural History) 1976-79." *Museums Journal* 80 (1980).

American Association for the Advancement of Science. "Chapter 13: Effective Learning and Teaching." *Science for All Americans*. Washington, DC: American Association for the Advancement of Science, 1989.

American Association of Museums. *Museums for a New Century*. Washington, DC: American Association of Museums, 1984.

Ames, Peter. "Measures of Merit?" *Museum News*, September/October 1991.

Anderson, Peter. *Before the Blueprint: Science Center Buildings*. Washington, DC: ASTC, 1991.

Association of Science-Technology Centers. *1990 ASTC Salary Survey*. Washington, DC: ASTC, 1990.

Barrier-Free in Brief. 4 vols. Washington, DC: American Association for the Advancement of Science, 1991.

Borun, Minda et al. *Planets and Pulleys: Studies of School Visits to Science Museums*. Washington, DC: ASTC, 1983.

Borun, Minda and Miller, Maryanne. "To Label or Not to Label?" *Museum News*, March/April 1980.

Bradburne, James. "Beyond Hands-On: Truth-telling and the Doing of Science." *The Nuffield Foundation Interactive Science and Technology Project Occasional Newsletter* 12 (July/August 1989).

Bruman, Raymond and Hipschman, Ron. *Exploratorium Cookbooks I, II, III*. San Francisco, CA: The Exploratorium, 1987.

Chabotar, Kent. "Cost Analysis in Schools and Other Nonprofits: A Management Perspective." *Urban Education* 24, no. 2 (July 1989).

Cox, D. "Attitudes to Science Among the Public Visiting Science Centres/Exhibitions." London: British Association for the Advancement of Science, unpublished project report.

Crane, H. Richard. *Explore & Discover: The Ann Arbor Hands-On Museum Exhibits Guide*. Ann Arbor: The Ann Arbor Hands-On Museum, 1991.

Csikszentmihalyi, Mihaly. "Human Behavior and the Science Center." In *Science Learning in the Informal Setting*, edited by Paul Heltne and Linda Marquardt. Chicago, IL: Chicago Academy of Sciences, 1988.

Danilov, Victor J. *Science and Technology Centers*. Cambridge, MA: The MIT Press, 1982.

Diamond, Judy et al. "The Exploratorium's Explainer Program: The Long-Term Impacts on Teenagers of Teaching Science to the Public." *Science Education* 71, no. 5 (1987).

Dimmock, Katherine. "Models of Adult Participation in Informal Science Education." Ph.D. dissertation, Northern Illinois University, 1985.

Doble, John and Richardson, Amy. "Scientific Issues and Thoughtful Public Involvement: A Case of the Impossible vs. the Inevitable?" New York: The Public Agenda Foundation, 1990.

Druger, Marvin, ed. *Science for the Fun of It: A Guide to Informal Science Education*. Washington, DC: National Science Teachers Association, 1988.

Duckworth, Eleanor. *The Having of Wonderful Ideas & Other Essays on Teaching and Learning*. New York: Teachers College, 1987.

Durant, John et al. "The Public Understanding of Science." *Nature*, 6 July 1989.

Educational Testing Service. *A World of Differences*. Princeton, NJ: Educational Testing Service, 1989.

Edward, Deborah et al. *Youth Volunteer Programs in Museums*. Austin, TX: Austin Children's Museum, 1989.

Evered, David and O'Connor, Maeve, eds. *Communicating Science to the Public*. New York: John Wiley and Sons, 1987.

Falk, Lisa. "'Not about Stuff, But for Somebody': Michael Spock on the Client-Centered Museum." *The Journal of Museum Education*, Fall 1987.

Feher, Elsa and Rice, Karen. "Development of Scientific Concepts Through the Use of Interactive Exhibits in a Museum." *Curator* 28, no. 1 (1985).

Flagg, Barbara. "Implementation Formative Evaluation of Earth Over Time Videodisc." Bellport, NY: Multimedia Research, 1990.

Gottfried, Jeffry. "Do Children Learn on School Field Trips?" *Curator* 23 (September 1980).

Great Explorations in Math and Science (GEMS) Series. Berkeley, CA: Lawrence Hall of Science.

Grinell, Sheila and Curlin, Patricia. *Using Scientist Volunteers at Museums*. Washington, DC: American Association for the Advancement of Science, 1990.

Hein, Hilde. *The Exploratorium: The Museum as Laboratory*. Washington, DC: Smithsonian Institution Press, 1990.

Heltne, Paul and Marquardt, Linda, eds. *Science Learning in the Informal Setting*. Chicago, IL: Chicago Academy of Sciences, 1988.

ILVS Bibliography and Abstracts. 3rd ed. Milwaukee, WI: International Laboratory for Visitor Studies, Department of Psychology, University of Wisconsin-Milwaukee, 1991.

Johnson, Roger and Johnson, David. "Cooperative Learning and the Achievement and Socialization Crises in Science and Mathematics Classrooms." In *Students and Science Learning*, edited by Audrey Champagne and Leslie Hornig. Washington, DC: American Association for the Advancement of Science, 1987.

Johnston, Douglas A. "The Law of Museum Safety." *The International Journal of Museum Management and Curatorship* 6 (1987).

Kennedy, Jeff. *User-Friendly: Hands-On Exhibits That Work*. Washington, DC: ASTC, 1990.

Knapp, Michael et al. *Opportunities for Strategic Investment in K-12 Science Education*. Palo Alto, CA: SRI International, 1987.

Levy, Shab. *Cogs, Cranks & Crates: Guidelines for Hands-On Traveling Exhibitions*. Washington, DC: ASTC, 1989.

Lewenstein, Bruce, ed. *When Science Meets the Public*. Washington, DC: American Association for the Advancement of Science, 1992.

Loomis, Ross. "Chapter 7: Using Evaluation to Improve Programs." *Museum Visitor Evaluation*, Nashville, TN: American Association for State and Local History, 1987.

———. *Museum Visitor Evaluation*. Nashville, TN: American Association for State and Local History, 1987.

Malcom, Shirley. "Who Will Do Science in the Next Century?" *Scientific American*, February 1990.

Marsh, Caryl. "Opening the Way for Questions." *Northeast Training News*, October 1980.

Maxwell, Margaret. "You Can Build a Campaign." *Fund Raising Management*, June 1989.

McCormick, Susan and Pollock, Wendy, eds. *The ASTC Science Center Survey Report Series: Exhibits Report and Directory, Education Report and Directory,* and *Administration and Finance Report*. Washington, DC: ASTC, 1988-1989.

McLean, Kathleen. *Planning for People in Museum Exhibitions*, Washington, DC: ASTC, forthcoming.

———, ed. *Recent and Recommended: A Museum Exhibition Bibliography with Notes from the Field*. Washington, DC: National Association for Museum Exhibition, 1991.

McManus, Paulette. "It's the Company You Keep...the Social Determination of Learning-related Behaviour in a Science Museum." *The International Journal of Museum Management and Curatorship* 6 (1987).

Miles, Roger. "Audiovisuals, a Suitable Case for Treatment." In *Visitor Studies: Theory, Research, & Practice*. Vol. 2, edited by Stephen Bitgood. Jacksonville, AL: The Center for Social Design, Jacksonville State University, 1989.

Miller, Jon. *The American People and Science Policy*. New York: Pergamon Press, 1983.

———. "Scientific Literacy." DeKalb: Public Opinion Laboratory, Northern Illinois University, 1989.

Morrison, Philip and Morrison, Phylis. "...But TV is Not Enough." *The AAAS Observer*, 3 November 1989.

National Science Board. *Science Indicators—1991*. Washington, DC: U.S. Government Printing Office, 1991 (updated biannually).

National Science Center for Communications and Electronics. *Estimates of Annual Number of Visitors to the NSCCE Discovery Center: Executive Report*. Fort Gordon, GA: National Science Center for Communications and Electronics, 1989.

Nichols, David. "Managing the Transition into the Campaign." *Fund Raising Management*, June 1989.

Nichols, Susan, ed. *Organizing Your Museum: The Essentials*. Washington, DC: American Association of Museums Technical Information Service, 1989.

Office of Technology Assessment. *Educating Scientists and Engineers: Grade School to Grad School*. Washington, DC: U.S. Government Printing Office # 052-003-01110-7, 1988.

Oppenheimer, Frank et al. *Working Prototypes*. San Francisco: The Exploratorium, 1986.

Pitman-Gelles, Bonnie. *Museums, Magic & Children: Youth Education in Museums*. Washington, DC: ASTC, 1982.

Robinson, Michael. "Bioscience Education through Bioparks." *BioScience* 38, no. 9.

The Royal Society. "The Public Understanding of Science." London: The Royal Society, 1985.

Serrell, Beverly. *Making Exhibit Labels: A Step by Step Guide*. Nashville, TN: American Association for State and Local History, 1983.

———, ed. *What Research Says about Learning in Science Museums*. Washington DC: ASTC, 1990.

Shaver, Carl. "The Rights and Rituals of Fund Raising." *Museum News*, February 1973.

Shortland, Michael. "No Business Like Show Business." *Nature* 328 (1987).

Solomon, Joan. *Teaching Children in the Laboratory*. London: Croom Helm, 1980.

St. John, Mark. *First Hand Learning: Teacher Education in Science Museums*. 2 vols. Washington, DC: ASTC, 1990.

———. "New Metaphors for Carrying Out Evaluations in the Science Museum Setting." *Visitor Behavior*, Fall 1990.

St. John, Mark and Grinell, Sheila. *The ASTC Science Center Survey: An Independent Review of Findings*. Washington, DC: ASTC, 1989.

Sullivan, Robert. "Trouble in Paradigms." *Museum News*, January/February 1992.

Taylor, Samuel. *Try It! Improving Exhibits through Formative Evaluation*. Washington, DC: ASTC, 1991.

Ullberg, Alan and Ullberg, Patricia. *Museum Trusteeship*. Washington, DC: American Association of Museums, 1981.

Weber, Joseph. *Managing the Board of Directors*. New York: The Greater New York Fund, Inc., 1975.

Weil, Stephen E. "A Checklist of Legal Considerations for Museums." *Museum News*, September/October 1987.

Weiss, Iris. *Report of the 1985-86 National Survey of Science and Mathematics Education*. Research Triangle Park, NC: Research Triangle Institute, 1987.

Wilcoxon, Sandra. "Measuring Your Impact." *Museum News*, November/December 1991.

Yager, Robert E. "The Constructivist Learning Model: Towards Real Reform in Science Education." *The Science Teacher*, September 1991.

Additional information is available through the Office of Museum Programs at the Smithsonian Institution in Washington, DC, which publishes a selection of bibliographies on various aspects of museum practice.

WORKS ON AUDIENCE RESEARCH SUGGESTED BY MARILYN HOOD:

Bigman, Stanley K. "Art Exhibit Audiences: Selected Findings on Who Comes?" *Museologist*, no. 60 (September 1956): 2-6.

Bishop, D.W., and Witt, P.A. "Sources of Behavioral Variance during Leisure Time." *Journal of Personality and Social Psychology* 16 (October 1970):352-60.

Cameron, Duncan F., and Abbey, David S. "Museum Audience Research." *Museum News* 40, no. 2 (October 1961):34-38.

Cheek, Neil H., Jr. and Burch, Jr., William R. *The Social Organization of Leisure in Human Society*. New York: Harper and Row, 1976.

Dillman, Don A. *Mail and Telephone Surveys: The Total Design Method*. New York: John Wiley and Sons, 1978.

DiMaggio, Paul, Useem, Michael, and Brown, Paula. *Audience Studies of the Performing Arts and Museums: A Critical Review*. Washington DC: National Endowment for the Arts, 1978.

Dixon, Brian, Courtney, Alice E., and Bailey, Robert H. *The Museum and the Canadian Public*. Toronto: Culturcan Publications, 1974.

Gudykunst, W.B.; Morra, J.A.; Kantor, W.I.; and Parker, H.A. "Dimensions of Leisure Activities: A Factor Analytic Study in New England." *Journal of Leisure Research* 13, no. 1 (1981):28-42.

Hawes, Douglass K., Talarzyk, W. Wayne, and Blackwell, Roger D. "Consumer Satisfactions from Leisure Time Pursuits." *Advances in Consumer Research* 2 (1975): 817-36.

Hendon, William S. *Analyzing an Art Museum*. New York: Praeger Publishing, 1979.

Iso-Ahola, Seppo E. "Some Social Psychological Determinants of Perceptions of Leisure: Preliminary Evidence." *Leisure Sciences* 2, no. 3/4 (1979):305-14.

Kelly, John R. "Leisure Socialization: Replication and Extension." *Journal of Leisure Research* 9, no. 2 (1977):121-32.

Morris, Rudolph E. "Leisure Time and the Museum." *Museum News* 41, no. 4 (December 1962):17-21.

Nash, George. "Art Museums as Perceived by the Public." *Curator* 18 (March 1975):55-67.

National Research Center of the Arts. *Americans and the Arts*. New York: American Council for the Arts, 1981.

Pommerehne, Werner W., and Frey, Bruno S. "The Museum from an Economic Perspective." *International Social Science Journal* 32, no. 2 (1980):323-39.

Robbins, J. E., and Robbins, S.S. "Museum Marketing: Identification of High, Moderate, and Low Attendee Segments." *Journal of the Academy of Marketing Science* 9, no. 1 (Winter 1981):66-76.

Wells, William D., ed. *Life Style and Psychographics*. Chicago, IL: American Marketing Association, 1974.

Yoesting, Dean R., and Christensen, James E. "Reexamining the Significance of Childhood Recreation Patterns on Adult Leisure Behavior." *Leisure Sciences* 1, no. 3 (1978):219-29.

WORKS ON COGNITIVE PSYCHOLOGY SUGGESTED BY LAUREN RESNICK AND MICHELENE CHI:

Bransford, J.D.; Stein, B.S.; Vye, N.J.; Franks, J.J.; Auble, P.M.; Mezynski, K.J.; and Perfetto, G.A. "Differences in Approaches to Learning: An Overview." *Journal of Experimental Psychology: General, III*, 1982: 390-398.

Carey, S. "Are Children Fundamentally Different Kinds of Thinkers and Learners than Adults?" In *Thinking and Learning Skills: Vol. 2. Research and Open Questions*, edited by S.F. Chipman, J. Siegel, and R. Glaser. Hillsdale, NJ: Erlbaum, 1985: 485-517.

Chi, M.T.H.; Bassok, M.; Lewis, M.; Reimann, P.; and Glaser, R. *Self Explanations: How Students Study and Use Examples in Learning to Solve Problems*, forthcoming.

Gelman, R. and Baillargcon, R. "A Review of Some Piagetian Concepts." In *Handbook of Child Psychology: Vol III, Cognitive Development*, edited by J.H. Flavell and E.M. Markman. New York: John Wiley and Sons, 1983: 167-230.

Gobbo, C., and Chi, M.T.H. "How Knowledge is Structured and Used by Expert and Novice Children." *Cognitive Development*, 1968: 221-237.

Piaget, J. *The Child's Conception of Physical Causality*. London: Routledge and Kegan Paul, 1980.

————. *Understanding Causality*. New York: Norton, 1974.

Ranney, M. "Restructuring Naive Conceptions of Motion." Unpublished Ph.D. dissertation, University of Pittsburgh, 1987.

Vosniadou, S., and Brewer, W.F. *Theories of Knowledge Restructuring in Development*. Review of Educational Research 57 (1987): 51-68.

White, B., and Horowitz, P. "Thinker Tools: Causal Models, Conccptual Change, and Science Education." *Cognition and Instruction*, in press.